Workbook to accompany

MaterialTool™
A Selection Guide of Materials and Processes for Designers

Andrzej Jan Wroblewski
Suryanarayana Vanka

Prentice Hall
Upper Saddle River, New Jersey   Columbus, Ohio

Editor: Stephen Helba
Assistant Editor: Michelle Churma
Design Coordinator: Karrie M. Converse-Jones
Production Manager: Deidra M. Schwartz
Marketing Manager: Chris Bracken

© 1999 by Prentice-Hall, Inc.
Pearson Education
Upper Saddle River, New Jersey 07458

All rights reserved. No part of this book may be reproduced, in any form or by any means, without permission in writing from the publisher.

Printed in the United States of America

10 9 8 7 6 5 4 3 2 1

ISBN: 0-13-011079-5

Prentice-Hall International (UK) Limited, *London*
Prentice-Hall of Australia Pty. Limited, *Sydney*
Prentice-Hall of Canada, Inc., *Toronto*
Prentice-Hall Hispanoamericana, S. A., *Mexico*
Prentice-Hall of India Private Limited, *New Delhi*
Prentice-Hall of Japan, Inc., *Tokyo*
Simon & Schuster Asia Pte. Ltd., *Singapore*
Editora Prentice-Hall do Brasil, Ltda., *Rio de Janeiro*

CONTENTS

Introduction	....	2
Unit 1. Pastry Container	....	14
Unit 2. Sake Box	....	25
Unit 3. Stirrer	....	36
Unit 4. Table Cloth	....	47
Unit 5. Storage Box	....	58
Unit 6. Screen Construction Block	....	69
Unit 7. Teapot	....	80
Unit 8. Nail Clipper	....	91
Unit 9. Belt	....	102
Unit 10. Garden Tool	....	113
Unit 11. Salt & Pepper Shakers	....	124
Unit 12. Bottle Opener	....	135
Unit 13. Garlic Press	....	146
Unit 14. Baby Bottle	....	157
Unit 15. Coffee Pot	....	168
Unit 16. Clothes Clip	....	179
Unit 17. Grater	....	190
Unit 18. Book	....	201
Unit 19. Lobster Cracker	....	212
Unit 20. Cement Edger	....	223
Solutions	....	234

INTRODUCTION

What is MaterialTool?

MaterialTool: A Selection Guide of Materials and Processes is a software design tool that assists designers in developing products for mass-production. MaterialTool catalogs hundreds of materials and manufacturing processes, and assists designers in selecting those appropriate to each project. Using MaterialTool, designers can plan and predict the visual, functional, and commercial consequences of their decisions, and transform creative ideas into realistic products. MaterialTool is particularly suited to the needs of industrial designers, engineers, architects, technology educators, and design students.

How is MaterialTool used in design projects?

Each material and manufacturing process has unique capabilities and limitations which designers need a clear understanding of. Yet, keeping up with this deluge of ever-changing information and selecting among myriad possibilities can be quite overwhelming. MaterialTool distills this vast body of information, and catalogs knowledge that is most relevant to designers. This includes information on the physical and chemical properties, cost considerations, and aesthetic possibilities of each material, and the form geometrics, surface attributes, and production capabilities of each process. MaterialTool presents designers with appropriate information, when they need it, in lively and accessible form so that their focus remains on what they do best: creative problem-solving.

How is MaterialTool used in learning?

MaterialTool can be used not only by practicing designers to solve problems, but also by students to learn about manufacturing. The MaterialTool Learning System consists of three components: MaterialTool, an Instructor's Manual, and a Student Workbook.

- **MaterialTool**: Instructors of structured materials and processes curricula use MaterialTool as a lecture aid. Students use MaterialTool outside the classroom as a self-paced learning guide, to refine their understanding of the lectures through self-directed review. Students also use MaterialTool as a reference tool in assigned design projects.
- **Instructor's Manual**: Instructors use the Instructor's Manual to structure course curricula and create handouts for students. Students receive copies of appropriate handouts from the Instructor Manual when appropriate.
- **Student Workbook**: Instructors assign students exercises from the accompanying Student Workbook. In these exercises, students are challenged to find the best possible manufacturing solutions for a number of product scenarios.

This Student Workbook

This Student Workbook accompanies "MaterialTool: A Selection Guide of Materials and Processes."

- The Workbook consists of 20 exercises, which take the student through an instructional sequence of materials and manufacturing processes.
- Each unit is identical in structure. The student is first presented with a photograph of a manufactured product, and a scenario surrounding its manufacture. The student is challenged to discover the materials the product is made from, and the processes used to bring it to its fully manufactured form. A structured series of steps guides the student through this process of discovery.
- The use of MaterialTool is integrated into the very fabric of this discovery process, which allows the student to learn MaterialTool as they learn the instructional material. On completion of the sequence of exercises the student thus not only gains an understanding of manufacturing, but also expertise in using MaterialTool.
- Solutions for all exercises are provided within the Workbook so that students can independently complete exercises outside of class, and motivated students can rapidly progress through self-directed learning.

Example

Unit X: Kitchen Paring Knife

Problem
The picture on the right shows a kitchen paring knife. This knife was designed specifically with three qualities in mind: durability, ease of use, and high standard of quality. Based on the provided picture, what advice can you offer regarding the choice of materials and manufacturing processes?
Keep in mind that this product needs to have a high aesthetic value and good performance.

SELECTING MATERIALS

Step 1: Identify the Parts of the Product
The product shown above may be made up of more than one part. Examine the photograph of the product, and identify all the parts of the product. List these in Worksheet 1: List of all Product Parts.

1.	Blade
2.	Handle: Left and Right Halves
3.	Rivets
4.	
5.	
6.	
7.	
8.	
9.	
10.	

Worksheet 1: List of all Product Parts

Step 2: List Unique Parts
Some of the parts listed above may be identical parts. Identify the unique parts of the product. Consider mirrored parts, such as a left half and a right half of a product, to be identical parts. List up to three unique parts in Worksheet 2: Unique Product Parts.

A	Blade
B	Handle
C	Rivet

Worksheet 2: Unique Product Parts

3

Step 3: Define the Desired Property Values

First identify what performance characteristics are required for each part that has been identified above. Circle the property values for the material that satisfy the performance requirements for each part in Worksheet 3: Property Values below. Note that only some of the properties will be relevant for each part; leave the rest blank.

	Part A						
Hardness		Low		Medium		**High**	
Impact Strength		Low		Medium		**High**	
Rigidity		Low		**Medium**		High	
Abrasion Resistance		Low		Medium		High	
Optical Clarity		Low		Medium		High	
Chemical Resistance		Low		**Medium**		High	
Food Compatibility		Low		Medium		**High**	
Ultraviolet Stability		Low		Medium		High	
Moisture Resistance		Low		Medium		**High**	
Lubricity		Low		Medium		High	
Electrical Insulation		Low		Medium		High	
Heat Resistance		Low		Medium		High	
Recyclability		Low		Medium		High	
Degradability		Low		Medium		High	
Cost		Low		**Medium**		**High**	
Color	Black	Gray	**White**	Blue	Green	Red	Yellow
	Part B						
Hardness		Low		Medium		**High**	
Impact Strength		Low		Medium		**High**	
Rigidity		Low		Medium		**High**	
Abrasion Resistance		Low		Medium		High	
Optical Clarity		Low		Medium		High	
Chemical Resistance		Low		Medium		High	
Food Compatibility		Low		Medium		**High**	
Ultraviolet Stability		Low		Medium		High	
Moisture Resistance		Low		Medium		**High**	
Lubricity		Low		Medium		High	
Electrical Insulation		Low		Medium		High	
Heat Resistance		Low		Medium		High	
Recyclability		Low		Medium		High	
Degradability		Low		Medium		High	
Cost		Low		**Medium**		**High**	
Color	**Black**	Gray	White	Blue	Green	**Red**	Yellow
	Part C						
Hardness		Low		**Medium**		High	
Impact Strength		**Low**		**Medium**		High	
Rigidity		Low		**Medium**		High	
Abrasion Resistance		Low		Medium		High	
Optical Clarity		Low		Medium		High	
Chemical Resistance		Low		Medium		High	
Food Compatibility		Low		Medium		**High**	
Ultraviolet Stability		Low		Medium		High	
Moisture Resistance		Low		Medium		**High**	
Lubricity		Low		Medium		High	
Electrical Insulation		Low		Medium		High	
Heat Resistance		Low		Medium		High	
Recyclability		Low		Medium		High	
Degradability		Low		Medium		High	
Cost		Low		**Medium**		High	
Color	Black	Gray	White	Blue	Green	Red	**Yellow**

Worksheet 3: Property Values

Step 4: Determine Matching Material Families

On the MaterialTool Title page click on the button called "Materials." This takes you to the Materials Section. Click on the large button called "Properties." This takes you to the Materials Properties page. Notice the list of material families listed on the left: Ceramic, Cement, Fiber, and so on. In the right pane you see a list of properties such as Hardness, Impact Strength, Rigidity, and so on. Click on the values Low, Medium, or High to determine Material Families that match the choices from Worksheet 3: Property Values. List the matching families for Part A in Worksheet 4a, the matching families for Part B in Worksheet 4b, and the matching families for Part C in Worksheet 4c.

Part A				
Properties	**Value Selected**			**Matching Materials**
Hardness	Low	Medium	**High**	Ceramic, Cement, Fiber, Glass, Leather, Metal, Plastic, Stone, Wood.
Impact Strength	Low	Medium	**High**	Ceramic, Cement, Fiber, Glass, Leather, Metal, Plastic, Stone, Wood.
Rigidity	Low	**Medium**	High	Fiber, Glass, Leather, Metal, Plastic, Stone, Wood.
Abrasion Resistance	Low	Medium	**High**	Fiber, Glass, Leather, Metal, Plastic, Stone, Wood.
Optical Clarity	Low	Medium	High	
Chemical Resistance	Low	Medium	High	
Food Compatibility	Low	Medium	**High**	Ceramic, Fiber, Glass, Metal, Plastic, Stone, Wood.
Ultraviolet Stability	Low	Medium	High	
Moisture Resistance	Low	Medium	**High**	Ceramic, Fiber, Glass, Metal, Plastic, Stone, Wood.
Lubricity	Low	Medium	High	
Electrical Insulation	Low	Medium	High	
Heat Resistance	Low	Medium	High	
Recyclability	Low	Medium	High	
Degradability	Low	Medium	High	
Cost	Low	Medium	**High**	Fiber, Glass, Metal, Plastic, Stone, Wood.
Color		Black		
		Gray		
		White		Fiber, Glass, Metal, Plastic, Stone, Wood.
		Blue		
		Green		
		Red		
		Yellow		

Worksheet 4a: Matching Material Families

Step 5: Identify the Most Appropriate Material Families

List up to three material families that satisfy all the requirements listed in Worksheet 4a.

Part A					
1.	Metal	2.		3.	Plastic

Worksheet 5a: List of Matching Material Families

Part B				
Properties	**Value Selected**			**Matching Materials**
Hardness	Low	Medium	**High**	Ceramic, Cement, Fiber, Glass, Leather, Metal, Plastic, Stone, Wood.
Impact Strength	Low	Medium	**High**	Ceramic, Cement, Fiber, Glass, Leather, Metal, Plastic, Stone, Wood.
Rigidity	Low	Medium	**High**	Ceramic, Cement, Fiber, Glass, Leather, Metal, Plastic, Stone, Wood.
Abrasion Resistance	Low	Medium	**High**	Ceramic, Cement, Fiber, Glass, Leather, Metal, Plastic, Stone, Wood.
Optical Clarity	Low	Medium	High	
Chemical Resistance	Low	Medium	High	
Food Compatibility	Low	Medium	**High**	Ceramic, Fiber, Glass, Metal, Plastic, Stone, Wood.
Ultraviolet Stability	Low	Medium	High	
Moisture Resistance	Low	Medium	**High**	Ceramic, Fiber, Glass, Metal, Plastic, Stone, Wood.
Lubricity	Low	Medium	High	
Electrical Insulation	Low	Medium	High	
Heat Resistance	Low	Medium	High	
Recyclability	Low	Medium	High	
Degradability	Low	Medium	High	
Cost	Low	Medium	**High**	Ceramic, Fiber, Glass, Metal, Plastic, Stone, Wood.
Color	**Black**			Ceramic, Fiber, Glass, Metal, Plastic, Stone, Wood.
	Gray			
	White			
	Blue			
	Green			
	Red			Ceramic, Fiber, Glass, Metal, Plastic, Stone, Wood.
	Yellow			

Worksheet 4b: Matching Material Families

List up to three material families that satisfy all the requirements listed in Worksheet 4b.

Part B					
1.	Wood	2.	Plastic	3.	Ceramic

Worksheet 5b: List of Matching Material Families

Part C				
Properties	**Value Selected**			**Matching Materials**
Hardness	Low	**Medium**	High	Ceramic, Fiber, Glass, Leather, Metal, Paper, Plastic, Rubber, Stone, Wood.
Impact Strength	Low	**Medium**	High	Ceramic, Fiber, Glass, Leather, Metal, Paper, Plastic, Rubber, Stone, Wood.
Rigidity	Low	**Medium**	High	Fiber, Glass, Leather, Metal, Paper, Plastic, Stone, Wood.
Abrasion Resistance	Low	**Medium**	High	Fiber, Glass, Leather, Metal, Paper, Plastic, Stone, Wood.
Optical Clarity	Low	Medium	High	
Chemical Resistance	Low	Medium	High	
Food Compatibility	Low	Medium	**High**	Fiber, Glass, Metal, Paper, Plastic, Stone, Wood.
Ultraviolet Stability	Low	Medium	High	
Moisture Resistance	Low	Medium	**High**	Fiber, Glass, Metal, Paper, Plastic, Stone, Wood.
Lubricity	Low	Medium	High	
Electrical Insulation	Low	Medium	High	
Heat Resistance	Low	Medium	High	
Recyclability	Low	Medium	High	
Degradability	Low	Medium	High	
Cost	Low	**Medium**	High	Fiber, Glass, Metal, Paper, Plastic, Stone, Wood.
Color	Black			
	Gray			
	White			Fiber, Glass, Metal, Paper, Plastic, Stone, Wood.
	Blue			
	Green			
	Red			
	Yellow			Fiber, Glass, Metal, Paper, Plastic, Stone, Wood.

Worksheet 4c: Matching Material Families

List up to three material families that satisfy all the requirements listed in Worksheet 4c.

Part C				
1.	Metal	2.	Plastic	3.

Worksheet 5c: List of Matching Material Families

Step 6: Determine Matching Materials

Once again, go to the Materials Section and click on the large button called "Properties." This takes you to the main Materials Properties page. Referring to the list you created in Worksheet 3, click the arrow next to the first family. This brings up detailed materials. In the right pane, you see a list of properties such as Hardness, Impact Strength, Rigidity, and so on. Once again select the materials that match the choices in Worksheet 1. List these below in Worksheet 6: Matching Materials. List the matching materials for Part A in Worksheet 6a, the matching materials for Part B in Worksheet 6b, and the matching materials for Part C in Worksheet 6c.

Part A				
Properties	**Value Selected**			**Matching Materials**
Hardness	Low	Medium	**High**	Alloy Steel, Carbon Steel, Iron, Stainless Steel, Tool Steel, Nickel, Titanium, Tungsten, Vanadium.
Impact Strength	Low	Medium	**High**	Alloy Steel, Carbon Steel, Stainless Steel, Tool Steel, Titanium, Vanadium.
Rigidity	Low	**Medium**	High	Alloy Steel, Carbon Steel, Stainless Steel.
Abrasion Resistance	Low	Medium	**High**	Alloy Steel, Carbon Steel, Stainless Steel.
Optical Clarity	Low	Medium	High	
Chemical Resistance	Low	Medium	High	
Food Compatibility	Low	Medium	**High**	Alloy Steel, Stainless Steel.
Ultraviolet Stability	Low	Medium	High	
Moisture Resistance	Low	Medium	**High**	Alloy Steel, Stainless Steel.
Lubricity	Low	Medium	High	
Electrical Insulation	Low	Medium	High	
Heat Resistance	Low	Medium	High	
Recyclability	Low	Medium	High	
Degradability	Low	Medium	High	
Cost	Low	Medium	**High**	Alloy Steel, Stainless Steel.
Color	Black			
	Gray			
	White			Alloy Steel, Stainless Steel.
	Blue			
	Green			
	Red			
	Yellow			

Worksheet 6a: Matching Materials

List up to three materials that best satisfy all the requirements listed in Worksheet 3.

Part A					
1.	Stainless Steel	2.	Alloy Steel	3.	

Worksheet 7a: List of Matching Materials

Part B				
Properties	**Value Selected**			**Matching Materials**
Hardness	Low	Medium	**High**	Ash, Birch, Ebony, Mahogany, Oak, Pearwood, Rosewood, Teak, Bamboo, Processed Wood.
Impact Strength	Low	Medium	**High**	Ash, Birch, Mahogany, Oak, Pearwood, Teak, Bamboo.
Rigidity	Low	Medium	**High**	Birch, Oak, Pearwood, Teak.
Abrasion Resistance	Low	Medium	**High**	
Optical Clarity	Low	Medium	High	
Chemical Resistance	Low	Medium	High	
Food Compatibility	Low	Medium	**High**	Birch, Oak, Pearwood, Teak.
Ultraviolet Stability	Low	Medium	High	
Moisture Resistance	Low	Medium	**High**	Teak.
Lubricity	Low	Medium	High	
Electrical Insulation	Low	Medium	High	
Heat Resistance	Low	Medium	High	
Recyclability	Low	Medium	High	
Degradability	Low	Medium	High	
Cost	Low	Medium	**High**	
Color		Black		
		Gray		
		White		
		Blue		
		Green		
		Red		Teak.
		Yellow		

Worksheet 6b: Matching Materials

List up to three materials that best satisfy all the requirements listed in Worksheet 3.

Part B					
1.	Teak	2.		3.	

Worksheet 7b: List of Matching Materials

Part C				
Properties	**Value Selected**			**Matching Materials**
Hardness	Low	**Medium**	High	Carbon Steel, Iron, Aluminum, Brass, Bronze, Copper, Magnesium.
Impact Strength	Low	**Medium**	High	Carbon Steel, Iron, Aluminum, Brass, Bronze, Magnesium.
Rigidity	Low	**Medium**	High	Carbon Steel, Iron, Aluminum, Brass, Bronze, Magnesium.
Abrasion Resistance	Low	**Medium**	High	Carbon Steel, Iron, Aluminum, Brass, Bronze, Magnesium.
Optical Clarity	Low	Medium	High	
Chemical Resistance	Low	Medium	High	
Food Compatibility	Low	Medium	**High**	Carbon Steel, Iron, Aluminum, Brass, Bronze, Magnesium.
Ultraviolet Stability	Low	Medium	High	
Moisture Resistance	Low	Medium	**High**	Carbon Steel, Aluminum, Brass, Bronze, Magnesium.
Lubricity	Low	Medium	High	
Electrical Insulation	Low	Medium	High	
Heat Resistance	Low	Medium	High	
Recyclability	Low	Medium	High	
Degradability	Low	Medium	High	
Cost	Low	**Medium**	High	Carbon Steel, Aluminum, Brass, Bronze, Magnesium.
Color		Black		
		Gray		
		White		Aluminum, Magnesium.
		Blue		
		Green		
		Red		
		Yellow		Brass, Bronze.

Worksheet 6c: Matching Materials

List up to three materials that best satisfy all the requirements listed in Worksheet 3.

Part C					
1.	Brass	2.	Bronze	3.	

Worksheet 7c: List of Matching Materials

SELECTING PROCESSES

Step 8: Define the Form Requirements
Study the problem description, define the form characteristics desired of the product, and list these below in Worksheet 8: List of Form Requirements.

Part A
This part should be flat and strong. It should be possible to sharpen it, and it should be capable of maintaining a sharp edge.

Part B
This part should be capable of taking on a three-dimensional volumetric form.

Part C
This part should be cylindrical, and be capable of some deformation under pressure.

Worksheet 8: List of Form Requirements

Step 9: Determine the Stages by which the Desired Form is Achieved
Determine how the form requirements, described in Worksheet 8: List of Form Requirements, can be achieved as a series of successive stages. List each part and the necessary operations required to achieve the desired form in stages. If there are multiple parts in the product, call them out as well. List this information below in Worksheet 9: Stages of Form Generation.

Stage	What form is achieved in this stage?
Part A	
1.	Correct quantity of material is formed into a rough flat shape.
2.	The same shape is squeezed into the precise shape of the blade.
3.	Any excess material is removed.
4.	Holes are made in the flat form.
5.	One side is sharpened to a fine edge; flat side is polished to a high finish.
Part B	
1.	Material is sized to approximate size.
2.	Material is shaped to correct form.
3.	Holes are made in the form.
Part C	
1.	Material is formed into a long cylinder.
2.	Cylinder is cut into small pieces.
3.	The small cylindrical pieces are used to assemble Parts A & B.
4.	Excess material is removed.

Worksheet 9: Stages of Form Generation

Step 10: Determine the Manufacturing Processes that are Relevant

Refer to the materials selections listed previously in Worksheet 7: List of Matching Materials, and the form requirements listed in Worksheet 9: Stages of Form Generation. Apply the procedure described below to determine the processes that are appropriate for each stage of processing each of the selected materials.

- Load MaterialTool and on the Title page, click on the button called "**Processes.**" This takes you to the Processes section of MaterialTool.
- Click on any of the process groups buttons: "**Finishing**", "**Forming**", "**Joining**" or "**Machining**."
  This takes you to the selected processes group page. Notice that all major material families (such as Ceramic, Leather, and Wood) are listed in the definition. By clicking on these words, it is possible to determine the processes applicable to each material group.
- Click on the button with the name of an individual process. This takes you to that particular process page. By reading the process definition, design tips, and visual consequences, determine if the process can be used to achieve the desired form in the particular material.
- For a deeper understanding of the process, click on the green button named "**Show principle.**"
  List the selected processes below in Worksheet 10: List of Selected Manufacturing Processes.

Step	Process	Description
Part A		
1.	Blocking	Steel workpiece is flattened to near final form.
2.	Closed Die Forging	The part is squeezed to its final thickness and form.
3.	Grinding	Excess flash is ground away and a clean sharp edge is created.
4.	Drilling	Holes are drilled for assembly.
5.	Polishing	The edge is further sharpened, and the surface is smoothened.
Part B		
1.	Circular Sawing	Rough form is created from wood.
2.	Milling	Final form is achieved by cutting away excess material.
3.	Drilling	Holes are drilled to accommodate the rivets.
4.		
5.		
Part C		
1.	Extruding	Brass or bronze is squeezed through a round die.
2.	Circular Sawing	The cylinder is cut into small cylindrical posts.
3.	Riveting	The cylindrical posts are used to assemble Parts A & B.
4.	Grinding	Excess material is ground away.
5.		

Worksheet 10: List of Selected Manufacturing Processes

FINAL SELECTION OF MATERIALS & PROCESSES

Step 11: List the Final Selections of Materials and Processes

Refer to the materials selections listed previously in Worksheet 7: List of Matching Materials, and the form requirements listed in Worksheet 10: List of Selected Manufacturing Processes. List the final choices of materials and manufacturing processes below in the Final Report.

	FINAL REPORT
	Part A
Material Selected	Alloy Steel or Stainless Steel.
Properties	Rustproof, hard, strong, food compatible.
Rationale	Stainless Steel or Alloy Steel creates a strong part that has a hard sharp edge and is resistant to rust.
Processes Selected	Forging, Grinding, Drilling, Polishing.
Operation	Shape the material into a flat piece, clean the forged edges, make holes for assembly, and create a polished surface.
Rationale	Although the part has minimal thickness, forging makes it very strong. The part needs to be ground to an accurate profile. The surface of the blade also has to be polished for functional and aesthetic reasons. Holes must be drilled into the part to make it possible to assemble it with the two halves of the handle.
	Part B
Material Selected	Teak.
Properties	Dense, waterproof, durable, rich red color.
Rationale	Teak is dense hardwood that is resistant to repeated washing and can bear impact of cutting.
Processes Selected	Circular Sawing, Milling, Drilling, Polishing.
Operation	Cut the material into the approximate size, shape it precisely into the desired profile, make holes for assembly, and smooth surface post assembly.
Rationale	Circular sawing is an inexpensive and fast way of achieving an approximate form. The complex form that is required of the handle is achieved by milling. Polishing creates a high quality surface that is free of any surface irregularities or burrs. Assembly requires precise holes that match the holes in the blade, a requirement which is achieved by drilling.
	Part C
Material Selected	Brass or Bronze.
Properties	Soft, attractive, resistant to water.
Rationale	Brass and Bronze are good riveting materials because they deform to make a tight fit. They do not rust or tarnish in the presence of water.
Processes Selected	Extrusion, Sawing, Riveting, Grinding.
Operation	Create a part with consistent circular profile, slice it into cylindrical components, mechanically squeeze component into the holes, and clean surface.
Rationale	Extrusion is an inexpensive way to create a continuous cylindrical form. Circular sawing is used to cut the length into the desired lengths that are used as rivets. Riveting is a quick and inexpensive way to create a strong permanent joint. Riveting does leave a minor amount of excess material outside the joint, which can be easily cleaned by a grinding process.

Worksheet 11: Final Report

Unit 1: Pastry Container

Problem

A manufacturer of plastic products wants to begin production of pastry containers, as shown in the picture. The containers need to be transparent, and only one kind of plastic can be used for this product. What kind of material should the manufacturer use to meet these requirements? What manufacturing processes should be selected and in what sequence should they be applied?

SELECTING MATERIALS

Step 1: Identify the Parts of the Product

The product shown above may be made up of more than one part. Examine the photograph of the product, and identify all the parts of the product. List these in Worksheet 1: List of all Product Parts.

1.	
2.	
3.	
4.	
5.	
6.	
7.	
8.	
9.	
10.	

Worksheet 1: List of all Product Parts

Step 2: List Unique Parts

Some of the parts listed above may be identical parts. Identify the unique parts of the product. Consider mirrored parts, such as a left half and a right half of a product, to be identical parts. List up to three unique parts in Worksheet 2: Unique Product Parts.

A	
B	
C	

Worksheet 2: Unique Product Parts

Step 3: Define the Desired Property Values

First identify what performance characteristics are required for each part that has been identified above. Circle the property values for the material that satisfy the performance requirements for each part in Worksheet 3: Property Values below. Note that only some of the properties will be relevant for each part; leave the rest blank.

Part A						
Hardness	Low		Medium		High	
Impact Strength	Low		Medium		High	
Rigidity	Low		Medium		High	
Abrasion Resistance	Low		Medium		High	
Optical Clarity	Low		Medium		High	
Chemical Resistance	Low		Medium		High	
Food Compatibility	Low		Medium		High	
Ultraviolet Stability	Low		Medium		High	
Moisture Resistance	Low		Medium		High	
Lubricity	Low		Medium		High	
Electrical Insulation	Low		Medium		High	
Heat Resistance	Low		Medium		High	
Recyclability	Low		Medium		High	
Degradability	Low		Medium		High	
Cost	Low		Medium		High	
Color	Black	Gray	White	Blue	Green	Red	Yellow

Part B						
Hardness	Low		Medium		High	
Impact Strength	Low		Medium		High	
Rigidity	Low		Medium		High	
Abrasion Resistance	Low		Medium		High	
Optical Clarity	Low		Medium		High	
Chemical Resistance	Low		Medium		High	
Food Compatibility	Low		Medium		High	
Ultraviolet Stability	Low		Medium		High	
Moisture Resistance	Low		Medium		High	
Lubricity	Low		Medium		High	
Electrical Insulation	Low		Medium		High	
Heat Resistance	Low		Medium		High	
Recyclability	Low		Medium		High	
Degradability	Low		Medium		High	
Cost	Low		Medium		High	
Color	Black	Gray	White	Blue	Green	Red	Yellow

Part C						
Hardness	Low		Medium		High	
Impact Strength	Low		Medium		High	
Rigidity	Low		Medium		High	
Abrasion Resistance	Low		Medium		High	
Optical Clarity	Low		Medium		High	
Chemical Resistance	Low		Medium		High	
Food Compatibility	Low		Medium		High	
Ultraviolet Stability	Low		Medium		High	
Moisture Resistance	Low		Medium		High	
Lubricity	Low		Medium		High	
Electrical Insulation	Low		Medium		High	
Heat Resistance	Low		Medium		High	
Recyclability	Low		Medium		High	
Degradability	Low		Medium		High	
Cost	Low		Medium		High	
Color	Black	Gray	White	Blue	Green	Red	Yellow

Worksheet 3: Property Values

Step 4: Determine Matching Material Families
On the MaterialTool Title page click on the button called "Materials." This takes you to the Materials Section. Click on the large button called "Properties." This takes you to the Materials Properties page. Notice the list of material families listed on the left: Ceramic, Cement, Fiber, and so on. In the right pane you see a list of properties such as Hardness, Impact Strength, Rigidity, and so on. Click on the values Low, Medium, or High to determine Material Families that match the choices from Worksheet 3: Property Values. List the matching families for Part A in Worksheet 4a, the matching families for Part B in Worksheet 4b, and the matching families for Part C in Worksheet 4c.

Part A				
Properties	**Value Selected**			**Matching Materials**
Hardness	Low	Medium	High	
Impact Strength	Low	Medium	High	
Rigidity	Low	Medium	High	
Abrasion Resistance	Low	Medium	High	
Optical Clarity	Low	Medium	High	
Chemical Resistance	Low	Medium	High	
Food Compatibility	Low	Medium	High	
Ultraviolet Stability	Low	Medium	High	
Moisture Resistance	Low	Medium	High	
Lubricity	Low	Medium	High	
Electrical Insulation	Low	Medium	High	
Heat Resistance	Low	Medium	High	
Recyclability	Low	Medium	High	
Degradability	Low	Medium	High	
Cost	Low	Medium	High	
Color	Black			
	Gray			
	White			
	Blue			
	Green			
	Red			
	Yellow			

Worksheet 4a: Matching Material Families

Step 5: Identify the Most Appropriate Material Families
List up to three material families that satisfy all the requirements listed in Worksheet 4a.

Part A					
1.		2.		3.	

Worksheet 5a: List of Matching Material Families

Part B				
Properties	Value Selected			Matching Materials
Hardness	Low	Medium	High	
Impact Strength	Low	Medium	High	
Rigidity	Low	Medium	High	
Abrasion Resistance	Low	Medium	High	
Optical Clarity	Low	Medium	High	
Chemical Resistance	Low	Medium	High	
Food Compatibility	Low	Medium	High	
Ultraviolet Stability	Low	Medium	High	
Moisture Resistance	Low	Medium	High	
Lubricity	Low	Medium	High	
Electrical Insulation	Low	Medium	High	
Heat Resistance	Low	Medium	High	
Recyclability	Low	Medium	High	
Degradability	Low	Medium	High	
Cost	Low	Medium	High	
Color	Black			
	Gray			
	White			
	Blue			
	Green			
	Red			
	Yellow			

Worksheet 4b: Matching Material Families

List up to three material families that satisfy all the requirements listed in Worksheet 4b.

Part B					
1.		2.		3.	

Worksheet 5b: List of Matching Material Families

Part C				
Properties	**Value Selected**			**Matching Materials**
Hardness	Low	Medium	High	
Impact Strength	Low	Medium	High	
Rigidity	Low	Medium	High	
Abrasion Resistance	Low	Medium	High	
Optical Clarity	Low	Medium	High	
Chemical Resistance	Low	Medium	High	
Food Compatibility	Low	Medium	High	
Ultraviolet Stability	Low	Medium	High	
Moisture Resistance	Low	Medium	High	
Lubricity	Low	Medium	High	
Electrical Insulation	Low	Medium	High	
Heat Resistance	Low	Medium	High	
Recyclability	Low	Medium	High	
Degradability	Low	Medium	High	
Cost	Low	Medium	High	
Color	Black			
	Gray			
	White			
	Blue			
	Green			
	Red			
	Yellow			

Worksheet 4c: Matching Material Families

List up to three material families that satisfy all the requirements listed in Worksheet 4c.

Part C		
1.	2.	3.

Worksheet 5c: List of Matching Material Families

Step 6: Determine Matching Materials
Once again, go to the Materials Section and click on the large button called "Properties." This takes you to the main Materials Properties page. Referring to the list you created in Worksheet 3, click the arrow next to the first family. This brings up detailed materials. In the right pane, you see a list of properties such as Hardness, Impact Strength, Rigidity, and so on. Once again select the materials that match the choices in Worksheet 1. List these below in Worksheet 6: Matching Materials. List the matching materials for Part A in Worksheet 6a, the matching materials for Part B in Worksheet 6b, and the matching materials for Part C in Worksheet 6c.

Part A				
Properties	**Value Selected**			**Matching Materials**
Hardness	Low	Medium	High	
Impact Strength	Low	Medium	High	
Rigidity	Low	Medium	High	
Abrasion Resistance	Low	Medium	High	
Optical Clarity	Low	Medium	High	
Chemical Resistance	Low	Medium	High	
Food Compatibility	Low	Medium	High	
Ultraviolet Stability	Low	Medium	High	
Moisture Resistance	Low	Medium	High	
Lubricity	Low	Medium	High	
Electrical Insulation	Low	Medium	High	
Heat Resistance	Low	Medium	High	
Recyclability	Low	Medium	High	
Degradability	Low	Medium	High	
Cost	Low	Medium	High	
Color	Black			
	Gray			
	White			
	Blue			
	Green			
	Red			
	Yellow			

Worksheet 6a: Matching Materials

List up to three materials that best satisfy all the requirements listed in Worksheet 3.

Part A					
1.		2.		3.	

Worksheet 7a: List of Matching Materials

Part B				
Properties	**Value Selected**			**Matching Materials**
Hardness	Low	Medium	High	
Impact Strength	Low	Medium	High	
Rigidity	Low	Medium	High	
Abrasion Resistance	Low	Medium	High	
Optical Clarity	Low	Medium	High	
Chemical Resistance	Low	Medium	High	
Food Compatibility	Low	Medium	High	
Ultraviolet Stability	Low	Medium	High	
Moisture Resistance	Low	Medium	High	
Lubricity	Low	Medium	High	
Electrical Insulation	Low	Medium	High	
Heat Resistance	Low	Medium	High	
Recyclability	Low	Medium	High	
Degradability	Low	Medium	High	
Cost	Low	Medium	High	
Color	Black			
	Gray			
	White			
	Blue			
	Green			
	Red			
	Yellow			

Worksheet 6b: Matching Materials

List up to three materials that best satisfy all the requirements listed in Worksheet 3.

Part B		
1.	2.	3.

Worksheet 7b: List of Matching Materials

Part C				
Properties	**Value Selected**			**Matching Materials**
Hardness	Low	Medium	High	
Impact Strength	Low	Medium	High	
Rigidity	Low	Medium	High	
Abrasion Resistance	Low	Medium	High	
Optical Clarity	Low	Medium	High	
Chemical Resistance	Low	Medium	High	
Food Compatibility	Low	Medium	High	
Ultraviolet Stability	Low	Medium	High	
Moisture Resistance	Low	Medium	High	
Lubricity	Low	Medium	High	
Electrical Insulation	Low	Medium	High	
Heat Resistance	Low	Medium	High	
Recyclability	Low	Medium	High	
Degradability	Low	Medium	High	
Cost	Low	Medium	High	
Color	Black			
	Gray			
	White			
	Blue			
	Green			
	Red			
	Yellow			

Worksheet 6c: Matching Materials

List up to three materials that best satisfy all the requirements listed in Worksheet 3.

Part C		
1.	2.	3.

Worksheet 7c: List of Matching Materials

SELECTING PROCESSES

Step 8: Define the Form Requirements
Study the problem description, define the form characteristics desired of the product, and list these below in Worksheet 8: List of Form Requirements.

Part A
Part B
Part C

Worksheet 8: List of Form Requirements

Step 9: Determine the Stages by which the Desired Form is Achieved
Determine how the form requirements, described in Worksheet 8: List of Form Requirements, can be achieved as a series of successive stages. List each part and the necessary operations required to achieve the desired form in stages. If there are multiple parts in the product, call them out as well. List this information below in Worksheet 9: Stages of Form Generation.

Stage	What form is achieved in this stage?
	Part A
	Part B
	Part C

Worksheet 9: Stages of Form Generation

Step 10: Determine the Manufacturing Processes that are Relevant

Refer to the materials selections listed previously in Worksheet 7: List of Matching Materials, and the form requirements listed in Worksheet 9: Stages of Form Generation. Apply the procedure described below to determine the processes that are appropriate for each stage of processing each of the selected materials.

- Load MaterialTool and on the Title page click on the button called "**Processes.**" This takes you to the Processes section of MaterialTool.
- Click on any of the process groups buttons: "**Finishing**", "**Forming**", "**Joining**" or "**Machining**."
  This takes you to the selected processes group page. Notice that all major material families (such as Ceramic, Leather, and Wood) are listed in the definition. By clicking on these words, it is possible to determine the processes applicable to each material group.
- Click on the button with the name of an individual process. This takes you to that particular process page. By reading the process definition, design tips, and visual consequences, determine if the process can be used to achieve the desired form in the particular material.
- For a deeper understanding of the process, click on the green button named "**Show principle**."
  List the selected processes below in Worksheet 10: List of Selected Manufacturing Processes.

Step	Process	Description
	Part A	
1.		
2.		
3.		
4.		
5.		
	Part B	
1.		
2.		
3.		
4.		
5.		
	Part C	
1.		
2.		
3.		
4.		
5.		

Worksheet 10: List of Selected Manufacturing Processes

FINAL SELECTION OF MATERIALS & PROCESSES

Step 11: List the Final Selections of Materials and Processes
Refer to the materials selections listed previously in Worksheet 7: List of Matching Materials, and the form requirements listed in Worksheet 10: List of Selected Manufacturing Processes. List the final choices of materials and manufacturing processes below in the Final Report.

FINAL REPORT	
Part A	
Material Selected	
Properties	
Rationale	
Processes Selected	
Operation	
Rationale	
Part B	
Material Selected	
Properties	
Rationale	
Processes Selected	
Operation	
Rationale	
Part C	
Material Selected	
Properties	
Rationale	
Processes Selected	
Operation	
Rationale	

Worksheet 11: Final Report

Unit 2: Sake Box

Problem
An exporter of small sake boxes is planning to open a new plant in your town. Based on the provided picture, what advice can you offer her regarding the choice of materials and manufacturing processes? Keep in mind that this product needs to have a high aesthetic value and has to be manufactured from natural materials.

SELECTING MATERIALS

Step 1: Identify the Parts of the Product
The product shown above may be made up of more than one part. Examine the photograph of the product, and identify all the parts of the product. List these in Worksheet 1: List of all Product Parts.

1.	
2.	
3.	
4.	
5.	
6.	
7.	
8.	
9.	
10.	

Worksheet 1: List of all Product Parts

Step 2: List Unique Parts
Some of the parts listed above may be identical parts. Identify the unique parts of the product. Consider mirrored parts, such as a left half and a right half of a product, to be identical parts. List up to three unique parts in Worksheet 2: Unique Product Parts.

A	
B	
C	

Worksheet 2: Unique Product Parts

Step 3: Define the Desired Property Values

First identify what performance characteristics are required for each part that has been identified above. Circle the property values for the material that satisfy the performance requirements for each part in Worksheet 3: Property Values below. Note that only some of the properties will be relevant for each part; leave the rest blank.

Part A							
Hardness	Low		Medium			High	
Impact Strength	Low		Medium			High	
Rigidity	Low		Medium			High	
Abrasion Resistance	Low		Medium			High	
Optical Clarity	Low		Medium			High	
Chemical Resistance	Low		Medium			High	
Food Compatibility	Low		Medium			High	
Ultraviolet Stability	Low		Medium			High	
Moisture Resistance	Low		Medium			High	
Lubricity	Low		Medium			High	
Electrical Insulation	Low		Medium			High	
Heat Resistance	Low		Medium			High	
Recyclability	Low		Medium			High	
Degradability	Low		Medium			High	
Cost	Low		Medium			High	
Color	Black	Gray	White	Blue	Green	Red	Yellow
Part B							
Hardness	Low		Medium			High	
Impact Strength	Low		Medium			High	
Rigidity	Low		Medium			High	
Abrasion Resistance	Low		Medium			High	
Optical Clarity	Low		Medium			High	
Chemical Resistance	Low		Medium			High	
Food Compatibility	Low		Medium			High	
Ultraviolet Stability	Low		Medium			High	
Moisture Resistance	Low		Medium			High	
Lubricity	Low		Medium			High	
Electrical Insulation	Low		Medium			High	
Heat Resistance	Low		Medium			High	
Recyclability	Low		Medium			High	
Degradability	Low		Medium			High	
Cost	Low		Medium			High	
Color	Black	Gray	White	Blue	Green	Red	Yellow
Part C							
Hardness	Low		Medium			High	
Impact Strength	Low		Medium			High	
Rigidity	Low		Medium			High	
Abrasion Resistance	Low		Medium			High	
Optical Clarity	Low		Medium			High	
Chemical Resistance	Low		Medium			High	
Food Compatibility	Low		Medium			High	
Ultraviolet Stability	Low		Medium			High	
Moisture Resistance	Low		Medium			High	
Lubricity	Low		Medium			High	
Electrical Insulation	Low		Medium			High	
Heat Resistance	Low		Medium			High	
Recyclability	Low		Medium			High	
Degradability	Low		Medium			High	
Cost	Low		Medium			High	
Color	Black	Gray	White	Blue	Green	Red	Yellow

Worksheet 3: Property Values

Step 4: Determine Matching Material Families

On the MaterialTool Title page click on the button called "Materials." This takes you to the Materials Section. Click on the large button called "Properties." This takes you to the Materials Properties page. Notice the list of material families listed on the left: Ceramic, Cement, Fiber, and so on. In the right pane you see a list of properties such as Hardness, Impact Strength, Rigidity, and so on. Click on the values Low, Medium, or High to determine Material Families that match the choices from Worksheet 3: Property Values. List the matching families for Part A in Worksheet 4a, the matching families for Part B in Worksheet 4b, and the matching families for Part C in Worksheet 4c.

Part A				
Properties	**Value Selected**			**Matching Materials**
Hardness	Low	Medium	High	
Impact Strength	Low	Medium	High	
Rigidity	Low	Medium	High	
Abrasion Resistance	Low	Medium	High	
Optical Clarity	Low	Medium	High	
Chemical Resistance	Low	Medium	High	
Food Compatibility	Low	Medium	High	
Ultraviolet Stability	Low	Medium	High	
Moisture Resistance	Low	Medium	High	
Lubricity	Low	Medium	High	
Electrical Insulation	Low	Medium	High	
Heat Resistance	Low	Medium	High	
Recyclability	Low	Medium	High	
Degradability	Low	Medium	High	
Cost	Low	Medium	High	
Color	Black			
	Gray			
	White			
	Blue			
	Green			
	Red			
	Yellow			

Worksheet 4a: Matching Material Families

Step 5: Identify the Most Appropriate Material Families

List up to three material families that satisfy all the requirements listed in Worksheet 4a.

Part A					
1.		2.		3.	

Worksheet 5a: List of Matching Material Families

Part B				
Properties	**Value Selected**			**Matching Materials**
Hardness	Low	Medium	High	
Impact Strength	Low	Medium	High	
Rigidity	Low	Medium	High	
Abrasion Resistance	Low	Medium	High	
Optical Clarity	Low	Medium	High	
Chemical Resistance	Low	Medium	High	
Food Compatibility	Low	Medium	High	
Ultraviolet Stability	Low	Medium	High	
Moisture Resistance	Low	Medium	High	
Lubricity	Low	Medium	High	
Electrical Insulation	Low	Medium	High	
Heat Resistance	Low	Medium	High	
Recyclability	Low	Medium	High	
Degradability	Low	Medium	High	
Cost	Low	Medium	High	
Color	Black			
	Gray			
	White			
	Blue			
	Green			
	Red			
	Yellow			

Worksheet 4b: Matching Material Families

List up to three material families that satisfy all the requirements listed in Worksheet 4b.

Part B					
1.		2.		3.	

Worksheet 5b: List of Matching Material Families

Part C				
Properties	**Value Selected**			**Matching Materials**
Hardness	Low	Medium	High	
Impact Strength	Low	Medium	High	
Rigidity	Low	Medium	High	
Abrasion Resistance	Low	Medium	High	
Optical Clarity	Low	Medium	High	
Chemical Resistance	Low	Medium	High	
Food Compatibility	Low	Medium	High	
Ultraviolet Stability	Low	Medium	High	
Moisture Resistance	Low	Medium	High	
Lubricity	Low	Medium	High	
Electrical Insulation	Low	Medium	High	
Heat Resistance	Low	Medium	High	
Recyclability	Low	Medium	High	
Degradability	Low	Medium	High	
Cost	Low	Medium	High	
Color	Black			
	Gray			
	White			
	Blue			
	Green			
	Red			
	Yellow			

Worksheet 4c: Matching Material Families

List up to three material families that satisfy all the requirements listed in Worksheet 4c.

Part C					
1.		2.		3.	

Worksheet 5c: List of Matching Material Families

Step 6: Determine Matching Materials
Once again, go to the Materials Section and click on the large button called "Properties." This takes you to the main Materials Properties page. Referring to the list you created in Worksheet 3, click the arrow next to the first family. This brings up detailed materials. In the right pane, you see a list of properties such as Hardness, Impact Strength, Rigidity, and so on. Once again select the materials that match the choices in Worksheet 1. List these below in Worksheet 6: Matching Materials. List the matching materials for Part A in Worksheet 6a, the matching materials for Part B in Worksheet 6b, and the matching materials for Part C in Worksheet 6c.

Part A				
Properties	**Value Selected**			**Matching Materials**
Hardness	Low	Medium	High	
Impact Strength	Low	Medium	High	
Rigidity	Low	Medium	High	
Abrasion Resistance	Low	Medium	High	
Optical Clarity	Low	Medium	High	
Chemical Resistance	Low	Medium	High	
Food Compatibility	Low	Medium	High	
Ultraviolet Stability	Low	Medium	High	
Moisture Resistance	Low	Medium	High	
Lubricity	Low	Medium	High	
Electrical Insulation	Low	Medium	High	
Heat Resistance	Low	Medium	High	
Recyclability	Low	Medium	High	
Degradability	Low	Medium	High	
Cost	Low	Medium	High	
Color	Black			
	Gray			
	White			
	Blue			
	Green			
	Red			
	Yellow			

Worksheet 6a: Matching Materials

List up to three materials that best satisfy all the requirements listed in Worksheet 3.

Part A		
1.	2.	3.

Worksheet 7a: List of Matching Materials

30

Part B				
Properties	**Value Selected**			**Matching Materials**
Hardness	Low	Medium	High	
Impact Strength	Low	Medium	High	
Rigidity	Low	Medium	High	
Abrasion Resistance	Low	Medium	High	
Optical Clarity	Low	Medium	High	
Chemical Resistance	Low	Medium	High	
Food Compatibility	Low	Medium	High	
Ultraviolet Stability	Low	Medium	High	
Moisture Resistance	Low	Medium	High	
Lubricity	Low	Medium	High	
Electrical Insulation	Low	Medium	High	
Heat Resistance	Low	Medium	High	
Recyclability	Low	Medium	High	
Degradability	Low	Medium	High	
Cost	Low	Medium	High	
Color	Black			
	Gray			
	White			
	Blue			
	Green			
	Red			
	Yellow			

Worksheet 6b: Matching Materials

List up to three materials that best satisfy all the requirements listed in Worksheet 3.

Part B					
1.		2.		3.	

Worksheet 7b: List of Matching Materials

Part C				
Properties	**Value Selected**			**Matching Materials**
Hardness	Low	Medium	High	
Impact Strength	Low	Medium	High	
Rigidity	Low	Medium	High	
Abrasion Resistance	Low	Medium	High	
Optical Clarity	Low	Medium	High	
Chemical Resistance	Low	Medium	High	
Food Compatibility	Low	Medium	High	
Ultraviolet Stability	Low	Medium	High	
Moisture Resistance	Low	Medium	High	
Lubricity	Low	Medium	High	
Electrical Insulation	Low	Medium	High	
Heat Resistance	Low	Medium	High	
Recyclability	Low	Medium	High	
Degradability	Low	Medium	High	
Cost	Low	Medium	High	
Color	Black			
	Gray			
	White			
	Blue			
	Green			
	Red			
	Yellow			

Worksheet 6c: Matching Materials

List up to three materials that best satisfy all the requirements listed in Worksheet 3.

Part C		
1.	2.	3.

Worksheet 7c: List of Matching Materials

SELECTING PROCESSES

Step 8: Define the Form Requirements
Study the problem description, define the form characteristics desired of the product, and list these below in Worksheet 8: List of Form Requirements.

Part A
Part B
Part C

Worksheet 8: List of Form Requirements

Step 9: Determine the Stages by which the Desired Form is Achieved
Determine how the form requirements, described in Worksheet 8: List of Form Requirements, can be achieved as a series of successive stages. List each part and the necessary operations required to achieve the desired form in stages. If there are multiple parts in the product, call them out as well. List this information below in Worksheet 9: Stages of Form Generation.

Stage	What form is achieved in this stage?
	Part A
	Part B
	Part C

Worksheet 9: Stages of Form Generation

33

Step 10: Determine the Manufacturing Processes that are Relevant
Refer to the materials selections listed previously in Worksheet 7: List of Matching Materials, and the form requirements listed in Worksheet 9: Stages of Form Generation. Apply the procedure described below to determine the processes that are appropriate for each stage of processing each of the selected materials.
- Load MaterialTool and on the Title page click on the button called "**Processes.**" This takes you to the Processes section of MaterialTool.
- Click on any of the process groups buttons: "**Finishing**", "**Forming**", "**Joining**" or "**Machining**."
  This takes you to the selected processes group page. Notice that all major material families (such as Ceramic, Leather, and Wood) are listed in the definition. By clicking on these words, it is possible to determine the processes applicable to each material group.
- Click on the button with the name of an individual process. This takes you to that particular process page. By reading the process definition, design tips, and visual consequences, determine if the process can be used to achieve the desired form in the particular material.
- For a deeper understanding of the process, click on the green button named "**Show principle**."
  List the selected processes below in Worksheet 10: List of Selected Manufacturing Processes.

Step	Process	Description
Part A		
1.		
2.		
3.		
4.		
5.		
Part B		
1.		
2.		
3.		
4.		
5.		
Part C		
1.		
2.		
3.		
4.		
5.		

Worksheet 10: List of Selected Manufacturing Processes

FINAL SELECTION OF MATERIALS & PROCESSES

Step 11: List the Final Selections of Materials and Processes
Refer to the materials selections listed previously in Worksheet 7: List of Matching Materials, and the form requirements listed in Worksheet 10: List of Selected Manufacturing Processes. List the final choices of materials and manufacturing processes below in the Final Report.

FINAL REPORT	
Part A	
Material Selected	
Properties	
Rationale	
Processes Selected	
Operation	
Rationale	
Part B	
Material Selected	
Properties	
Rationale	
Processes Selected	
Operation	
Rationale	
Part C	
Material Selected	
Properties	
Rationale	
Processes Selected	
Operation	
Rationale	

Worksheet 11: Final Report

Unit 3: Stirrer

Problem
One of your friends intends to open a small business manufacturing kitchen utensils. He wants to use only one kind of material and wants to ensure that the stirrer has a smooth surface. He plans to apply technology suitable for mass production, but is open to the possibility of some limited hand work as well. The manufacturer does not have the ability to manufacture items out of plastic.
What materials and technologies should he choose? In what sequence should the selected technological processes be applied?

SELECTING MATERIALS

Step 1: Identify the Parts of the Product
The product shown above may be made up of more than one part. Examine the photograph of the product, and identify all the parts of the product. List these in Worksheet 1: List of all Product Parts.

1.	
2.	
3.	
4.	
5.	
6.	
7.	
8.	
9.	
10.	

Worksheet 1: List of all Product Parts

Step 2: List Unique Parts
Some of the parts listed above may be identical parts. Identify the unique parts of the product. Consider mirrored parts, such as a left half and a right half of a product, to be identical parts. List up to three unique parts in Worksheet 2: Unique Product Parts.

Part A	
Part B	
Part C	

Worksheet 2: Unique Product Parts

Step 3: Define the Desired Property Values

First identify what performance characteristics are required for each part that has been identified above. Circle the property values for the material that satisfy the performance requirements for each part in Worksheet 3: Property Values below. Note that only some of the properties will be relevant for each part; leave the rest blank.

Part A						
Hardness	Low		Medium		High	
Impact Strength	Low		Medium		High	
Rigidity	Low		Medium		High	
Abrasion Resistance	Low		Medium		High	
Optical Clarity	Low		Medium		High	
Chemical Resistance	Low		Medium		High	
Food Compatibility	Low		Medium		High	
Ultraviolet Stability	Low		Medium		High	
Moisture Resistance	Low		Medium		High	
Lubricity	Low		Medium		High	
Electrical Insulation	Low		Medium		High	
Heat Resistance	Low		Medium		High	
Recyclability	Low		Medium		High	
Degradability	Low		Medium		High	
Cost	Low		Medium		High	
Color	Black	Gray	White	Blue	Green	Red	Yellow

Part B						
Hardness	Low		Medium		High	
Impact Strength	Low		Medium		High	
Rigidity	Low		Medium		**High**	
Abrasion Resistance	Low		Medium		**High**	
Optical Clarity	Low		Medium		**High**	
Chemical Resistance	Low		Medium		High	
Food Compatibility	Low		Medium		High	
Ultraviolet Stability	Low		Medium		High	
Moisture Resistance	Low		Medium		High	
Lubricity	Low		Medium		High	
Electrical Insulation	Low		Medium		High	
Heat Resistance	Low		Medium		High	
Recyclability	Low		Medium		High	
Degradability	Low		Medium		High	
Cost	Low		Medium		High	
Color	Black	Gray	White	Blue	Green	Red	Yellow

Part C						
Hardness	Low		Medium		High	
Impact Strength	Low		Medium		High	
Rigidity	Low		Medium		High	
Abrasion Resistance	Low		Medium		High	
Optical Clarity	Low		Medium		High	
Chemical Resistance	Low		Medium		High	
Food Compatibility	Low		Medium		High	
Ultraviolet Stability	Low		Medium		High	
Moisture Resistance	Low		Medium		High	
Lubricity	Low		Medium		High	
Electrical Insulation	Low		Medium		High	
Heat Resistance	Low		Medium		High	
Recyclability	Low		Medium		High	
Degradability	Low		Medium		High	
Cost	Low		Medium		High	
Color	Black	Gray	White	Blue	Green	Red	Yellow

Worksheet 3: Property Values

Step 4: Determine Matching Material Families

On the MaterialTool Title page click on the button called "Materials." This takes you to the Materials Section. Click on the large button called "Properties." This takes you to the Materials Properties page. Notice the list of material families listed on the left: Ceramic, Cement, Fiber, and so on. In the right pane you see a list of properties such as Hardness, Impact Strength, Rigidity, and so on. Click on the values Low, Medium, or High to determine Material Families that match the choices from Worksheet 3: Property Values. List the matching families for Part A in Worksheet 4a, the matching families for Part B in Worksheet 4b, and the matching families for Part C in Worksheet 4c.

Part A				
Properties	**Value Selected**			**Matching Materials**
Hardness	Low	Medium	High	
Impact Strength	Low	Medium	High	
Rigidity	Low	Medium	High	
Abrasion Resistance	Low	Medium	High	
Optical Clarity	Low	Medium	High	
Chemical Resistance	Low	Medium	High	
Food Compatibility	Low	Medium	High	
Ultraviolet Stability	Low	Medium	High	
Moisture Resistance	Low	Medium	High	
Lubricity	Low	Medium	High	
Electrical Insulation	Low	Medium	High	
Heat Resistance	Low	Medium	High	
Recyclability	Low	Medium	High	
Degradability	Low	Medium	High	
Cost	Low	Medium	High	
Color	Black			
	Gray			
	White			
	Blue			
	Green			
	Red			
	Yellow			

Worksheet 4a: Matching Material Families

Step 5: Identify the Most Appropriate Material Families

List up to three material families that satisfy all the requirements listed in Worksheet 4a.

Part A		
1.	2.	3.

Worksheet 5a: List of Matching Material Families

Part B				
Properties	**Value Selected**			**Matching Materials**
Hardness	Low	Medium	High	
Impact Strength	Low	Medium	High	
Rigidity	Low	Medium	High	
Abrasion Resistance	Low	Medium	High	
Optical Clarity	Low	Medium	High	
Chemical Resistance	Low	Medium	High	
Food Compatibility	Low	Medium	High	
Ultraviolet Stability	Low	Medium	High	
Moisture Resistance	Low	Medium	High	
Lubricity	Low	Medium	High	
Electrical Insulation	Low	Medium	High	
Heat Resistance	Low	Medium	High	
Recyclability	Low	Medium	High	
Degradability	Low	Medium	High	
Cost	Low	Medium	High	
Color	Black			
	Gray			
	White			
	Blue			
	Green			
	Red			
	Yellow			

Worksheet 4b: Matching Material Families

List up to three material families that satisfy all the requirements listed in Worksheet 4b.

Part B		
1.	2.	3.

Worksheet 5b: List of Matching Material Families

Part C				
Properties	**Value Selected**			**Matching Materials**
Hardness	Low	Medium	High	
Impact Strength	Low	Medium	High	
Rigidity	Low	Medium	High	
Abrasion Resistance	Low	Medium	High	
Optical Clarity	Low	Medium	High	
Chemical Resistance	Low	Medium	High	
Food Compatibility	Low	Medium	High	
Ultraviolet Stability	Low	Medium	High	
Moisture Resistance	Low	Medium	High	
Lubricity	Low	Medium	High	
Electrical Insulation	Low	Medium	High	
Heat Resistance	Low	Medium	High	
Recyclability	Low	Medium	High	
Degradability	Low	Medium	High	
Cost	Low	Medium	High	
Color	Black			
	Gray			
	White			
	Blue			
	Green			
	Red			
	Yellow			

Worksheet 4c: Matching Material Families

List up to three material families that satisfy all the requirements listed in Worksheet 4c.

Part C					
1.		2.		3.	

Worksheet 5c: List of Matching Material Families

Step 6: Determine Matching Materials
Once again, go to the Materials Section and click on the large button called "Properties." This takes you to the main Materials Properties page. Referring to the list you created in Worksheet 3, click the arrow next to the first family. This brings up detailed materials. In the right pane, you see a list of properties such as Hardness, Impact Strength, Rigidity, and so on. Once again select the materials that match the choices in Worksheet 1. List these below in Worksheet 6: Matching Materials. List the matching materials for Part A in Worksheet 6a, the matching materials for Part B in Worksheet 6b, and the matching materials for Part C in Worksheet 6c.

Properties	Value Selected			Matching Materials
		Part A		
Hardness	Low	Medium	High	
Impact Strength	Low	Medium	High	
Rigidity	Low	Medium	High	
Abrasion Resistance	Low	Medium	High	
Optical Clarity	Low	Medium	High	
Chemical Resistance	Low	Medium	High	
Food Compatibility	Low	Medium	High	
Ultraviolet Stability	Low	Medium	High	
Moisture Resistance	Low	Medium	High	
Lubricity	Low	Medium	High	
Electrical Insulation	Low	Medium	High	
Heat Resistance	Low	Medium	High	
Recyclability	Low	Medium	High	
Degradability	Low	Medium	High	
Cost	Low	Medium	High	
Color	Black			
	Gray			
	White			
	Blue			
	Green			
	Red			
	Yellow			

Worksheet 6a: Matching Materials

List up to three materials that best satisfy all the requirements listed in Worksheet 3.

Part A		
1.	2.	3.

Worksheet 7a: List of Matching Materials

41

Part B				
Properties	**Value Selected**			**Matching Materials**
Hardness	Low	Medium	High	
Impact Strength	Low	Medium	High	
Rigidity	Low	Medium	High	
Abrasion Resistance	Low	Medium	High	
Optical Clarity	Low	Medium	High	
Chemical Resistance	Low	Medium	High	
Food Compatibility	Low	Medium	High	
Ultraviolet Stability	Low	Medium	High	
Moisture Resistance	Low	Medium	High	
Lubricity	Low	Medium	High	
Electrical Insulation	Low	Medium	High	
Heat Resistance	Low	Medium	High	
Recyclability	Low	Medium	High	
Degradability	Low	Medium	High	
Cost	Low	Medium	High	
Color	Black			
	Gray			
	White			
	Blue			
	Green			
	Red			
	Yellow			

Worksheet 6b: Matching Materials

List up to three materials that best satisfy all the requirements listed in Worksheet 3.

Part B		
1.	2.	3.

Worksheet 7b: List of Matching Materials

Part C				
Properties	**Value Selected**			**Matching Materials**
Hardness	Low	Medium	High	
Impact Strength	Low	Medium	High	
Rigidity	Low	Medium	High	
Abrasion Resistance	Low	Medium	High	
Optical Clarity	Low	Medium	High	
Chemical Resistance	Low	Medium	High	
Food Compatibility	Low	Medium	High	
Ultraviolet Stability	Low	Medium	High	
Moisture Resistance	Low	Medium	High	
Lubricity	Low	Medium	High	
Electrical Insulation	Low	Medium	High	
Heat Resistance	Low	Medium	High	
Recyclability	Low	Medium	High	
Degradability	Low	Medium	High	
Cost	Low	Medium	High	
Color	Black			
	Gray			
	White			
	Blue			
	Green			
	Red			
	Yellow			

Worksheet 6c: Matching Materials

List up to three materials that best satisfy all the requirements listed in Worksheet 3.

Part C		
1.	2.	3.

Worksheet 7c: List of Matching Materials

SELECTING PROCESSES

Step 8: Define the Form Requirements
Study the problem description, define the form characteristics desired of the product, and list these below in Worksheet 8: List of Form Requirements.

Part A

Part B

Part C

Worksheet 8: List of Form Requirements

Step 9: Determine the Stages by which the Desired Form is Achieved
Determine how the form requirements, described in Worksheet 8: List of Form Requirements, can be achieved as a series of successive stages. List each part and the necessary operations required to achieve the desired form in stages. If there are multiple parts in the product, call them out as well. List this information below in Worksheet 9: Stages of Form Generation.

Stage	What form is achieved in this stage?
	Part A
	Part B
	Part C

Worksheet 9: Stages of Form Generation

Step 10: Determine the Manufacturing Processes that are Relevant

Refer to the materials selections listed previously in Worksheet 7: List of Matching Materials, and the form requirements listed in Worksheet 9: Stages of Form Generation. Apply the procedure described below to determine the processes that are appropriate for each stage of processing each of the selected materials.

- Load MaterialTool and on the Title page click on the button called "**Processes.**" This takes you to the Processes section of MaterialTool.
- Click on any of the process groups buttons: "**Finishing**", "**Forming**", "**Joining**" or "**Machining**."
  This takes you to the selected processes group page. Notice that all major material families (such as Ceramic, Leather, and Wood) are listed in the definition. By clicking on these words, it is possible to determine the processes applicable to each material group.
- Click on the button with the name of an individual process. This takes you to that particular process page. By reading the process definition, design tips, and visual consequences, determine if the process can be used to achieve the desired form in the particular material.
- For a deeper understanding of the process, click on the green button named "**Show principle**."
  List the selected processes below in Worksheet 10: List of Selected Manufacturing Processes.

Step	Process	Description
Part A		
1.		
2.		
3.		
4.		
5.		
Part B		
1.		
2.		
3.		
4.		
5.		
Part C		
1.		
2.		
3.		
4.		
5.		

Worksheet 10: List of Selected Manufacturing Processes

FINAL SELECTION OF MATERIALS & PROCESSES

Step 11: List the Final Selections of Materials and Processes
Refer to the materials selections listed previously in Worksheet 7: List of Matching Materials, and the form requirements listed in Worksheet 10: List of Selected Manufacturing Processes. List the final choices of materials and manufacturing processes below in the Final Report.

	FINAL REPORT
	Part A
Material Selected	
Properties	
Rationale	
Processes Selected	
Operation	
Rationale	
	Part B
Material Selected	
Properties	
Rationale	
Processes Selected	
Operation	
Rationale	
	Part C
Material Selected	
Properties	
Rationale	
Processes Selected	
Operation	
Rationale	

Worksheet 11: Final Report

Unit 4: Table Cloth

Problem
The picture on the right shows a table cloth. If you were to produce a similar product and were concerned about cost-effectiveness and aesthetic values, what specific materials and technologies would you select? Please describe all steps in the production process.

SELECTING MATERIALS

Step 1: Identify the Parts of the Product
The product shown above may be made up of more than one part. Examine the photograph of the product, and identify all the parts of the product. List these in Worksheet 1: List of all Product Parts.

1.	
2.	
3.	
4.	
5.	
6.	
7.	
8.	
9.	
10.	

Worksheet 1: List of all Product Parts

Step 2: List Unique Parts
Some of the parts listed above may be identical parts. Identify the unique parts of the product. Consider mirrored parts, such as a left half and a right half of a product, to be identical parts. List up to three unique parts in Worksheet 2: Unique Product Parts.

Part A	
Part B	
Part C	

Worksheet 2: Unique Product Parts

47

Step 3: Define the Desired Property Values

First identify what performance characteristics are required for each part that has been identified above. Circle the property values for the material that satisfy the performance requirements for each part in Worksheet 3: Property Values below. Note that only some of the properties will be relevant for each part; leave the rest blank.

	Part A						
Hardness	Low		Medium			High	
Impact Strength	Low		Medium			High	
Rigidity	Low		Medium			High	
Abrasion Resistance	Low		Medium			High	
Optical Clarity	Low		Medium			High	
Chemical Resistance	Low		Medium			High	
Food Compatibility	Low		Medium			High	
Ultraviolet Stability	Low		Medium			High	
Moisture Resistance	Low		Medium			High	
Lubricity	Low		Medium			High	
Electrical Insulation	Low		Medium			High	
Heat Resistance	Low		Medium			High	
Recyclability	Low		Medium			High	
Degradability	Low		Medium			High	
Cost	Low		Medium			High	
Color	Black	Gray	White	Blue	Green	Red	Yellow
	Part B						
Hardness	Low		Medium			High	
Impact Strength	Low		Medium			High	
Rigidity	Low		Medium			High	
Abrasion Resistance	Low		Medium			High	
Optical Clarity	Low		Medium			High	
Chemical Resistance	Low		Medium			High	
Food Compatibility	Low		Medium			High	
Ultraviolet Stability	Low		Medium			High	
Moisture Resistance	Low		Medium			High	
Lubricity	Low		Medium			High	
Electrical Insulation	Low		Medium			High	
Heat Resistance	Low		Medium			High	
Recyclability	Low		Medium			High	
Degradability	Low		Medium			High	
Cost	Low		Medium			High	
Color	Black	Gray	White	Blue	Green	Red	Yellow
	Part C						
Hardness	Low		Medium			High	
Impact Strength	Low		Medium			High	
Rigidity	Low		Medium			High	
Abrasion Resistance	Low		Medium			High	
Optical Clarity	Low		Medium			High	
Chemical Resistance	Low		Medium			High	
Food Compatibility	Low		Medium			High	
Ultraviolet Stability	Low		Medium			High	
Moisture Resistance	Low		Medium			High	
Lubricity	Low		Medium			High	
Electrical Insulation	Low		Medium			High	
Heat Resistance	Low		Medium			High	
Recyclability	Low		Medium			High	
Degradability	Low		Medium			High	
Cost	Low		Medium			High	
Color	Black	Gray	White	Blue	Green	Red	Yellow

Worksheet 3: Property Values

Step 4: Determine Matching Material Families

On the MaterialTool Title page click on the button called "Materials." This takes you to the Materials Section. Click on the large button called "Properties." This takes you to the Materials Properties page. Notice the list of material families listed on the left: Ceramic, Cement, Fiber, and so on. In the right pane you see a list of properties such as Hardness, Impact Strength, Rigidity, and so on. Click on the values Low, Medium, or High to determine Material Families that match the choices from Worksheet 3: Property Values. List the matching families for Part A in Worksheet 4a, the matching families for Part B in Worksheet 4b, and the matching families for Part C in Worksheet 4c.

Part A				
Properties	**Value Selected**			**Matching Materials**
Hardness	Low	Medium	High	
Impact Strength	Low	Medium	High	
Rigidity	Low	Medium	High	
Abrasion Resistance	Low	Medium	High	
Optical Clarity	Low	Medium	High	
Chemical Resistance	Low	Medium	High	
Food Compatibility	Low	Medium	High	
Ultraviolet Stability	Low	Medium	High	
Moisture Resistance	Low	Medium	High	
Lubricity	Low	Medium	High	
Electrical Insulation	Low	Medium	High	
Heat Resistance	Low	Medium	High	
Recyclability	Low	Medium	High	
Degradability	Low	Medium	High	
Cost	Low	Medium	High	
Color	Black			
	Gray			
	White			
	Blue			
	Green			
	Red			
	Yellow			

Worksheet 4a: Matching Material Families

Step 5: Identify the Most Appropriate Material Families

List up to three material families that satisfy all the requirements listed in Worksheet 4a.

Part A		
1.	2.	3.

Worksheet 5a: List of Matching Material Families

Part B				
Properties	**Value Selected**			**Matching Materials**
Hardness	Low	Medium	High	
Impact Strength	Low	Medium	High	
Rigidity	Low	Medium	High	
Abrasion Resistance	Low	Medium	High	
Optical Clarity	Low	Medium	High	
Chemical Resistance	Low	Medium	High	
Food Compatibility	Low	Medium	High	
Ultraviolet Stability	Low	Medium	High	
Moisture Resistance	Low	Medium	High	
Lubricity	Low	Medium	High	
Electrical Insulation	Low	Medium	High	
Heat Resistance	Low	Medium	High	
Recyclability	Low	Medium	High	
Degradability	Low	Medium	High	
Cost	Low	Medium	High	
Color	Black			
	Gray			
	White			
	Blue			
	Green			
	Red			
	Yellow			

Worksheet 4b: Matching Material Families

List up to three material families that satisfy all the requirements listed in Worksheet 4b.

Part B					
1.		2.		3.	

Worksheet 5b: List of Matching Material Families

Part C				
Properties	**Value Selected**			**Matching Materials**
Hardness	Low	Medium	High	
Impact Strength	Low	Medium	High	
Rigidity	Low	Medium	High	
Abrasion Resistance	Low	Medium	High	
Optical Clarity	Low	Medium	High	
Chemical Resistance	Low	Medium	High	
Food Compatibility	Low	Medium	High	
Ultraviolet Stability	Low	Medium	High	
Moisture Resistance	Low	Medium	High	
Lubricity	Low	Medium	High	
Electrical Insulation	Low	Medium	High	
Heat Resistance	Low	Medium	High	
Recyclability	Low	Medium	High	
Degradability	Low	Medium	High	
Cost	Low	Medium	High	
Color	Black			
	Gray			
	White			
	Blue			
	Green			
	Red			
	Yellow			

Worksheet 4c: Matching Material Families

List up to three material families that satisfy all the requirements listed in Worksheet 4c.

Part C					
1.		2.		3.	

Worksheet 5c: List of Matching Material Families

Step 6: Determine Matching Materials

Once again, go to the Materials Section and click on the large button called "Properties." This takes you to the main Materials Properties page. Referring to the list you created in Worksheet 3, click the arrow next to the first family. This brings up detailed materials. In the right pane, you see a list of properties such as Hardness, Impact Strength, Rigidity, and so on. Once again select the materials that match the choices in Worksheet 1. List these below in Worksheet 6: Matching Materials. List the matching materials for Part A in Worksheet 6a, the matching materials for Part B in Worksheet 6b, and the matching materials for Part C in Worksheet 6c.

Part A				
Properties	**Value Selected**			**Matching Materials**
Hardness	Low	Medium	High	
Impact Strength	Low	Medium	High	
Rigidity	Low	Medium	High	
Abrasion Resistance	Low	Medium	High	
Optical Clarity	Low	Medium	High	
Chemical Resistance	Low	Medium	High	
Food Compatibility	Low	Medium	High	
Ultraviolet Stability	Low	Medium	High	
Moisture Resistance	Low	Medium	High	
Lubricity	Low	Medium	High	
Electrical Insulation	Low	Medium	High	
Heat Resistance	Low	Medium	High	
Recyclability	Low	Medium	High	
Degradability	Low	Medium	High	
Cost	Low	Medium	High	
Color	Black			
	Gray			
	White			
	Blue			
	Green			
	Red			
	Yellow			

Worksheet 6a: Matching Materials

List up to three materials that best satisfy all the requirements listed in Worksheet 3.

Part A		
1.	2.	3.

Worksheet 7a: List of Matching Materials

Part B				
Properties	**Value Selected**			**Matching Materials**
Hardness	Low	Medium	High	
Impact Strength	Low	Medium	High	
Rigidity	Low	Medium	High	
Abrasion Resistance	Low	Medium	High	
Optical Clarity	Low	Medium	High	
Chemical Resistance	Low	Medium	High	
Food Compatibility	Low	Medium	High	
Ultraviolet Stability	Low	Medium	High	
Moisture Resistance	Low	Medium	High	
Lubricity	Low	Medium	High	
Electrical Insulation	Low	Medium	High	
Heat Resistance	Low	Medium	High	
Recyclability	Low	Medium	High	
Degradability	Low	Medium	High	
Cost	Low	Medium	High	
Color	Black			
	Gray			
	White			
	Blue			
	Green			
	Red			
	Yellow			

Worksheet 6b: Matching Materials

List up to three materials that best satisfy all the requirements listed in Worksheet 3.

Part B		
1.	2.	3.

Worksheet 7b: List of Matching Materials

Part C				
Properties	**Value Selected**			**Matching Materials**
Hardness	Low	Medium	High	
Impact Strength	Low	Medium	High	
Rigidity	Low	Medium	High	
Abrasion Resistance	Low	Medium	High	
Optical Clarity	Low	Medium	High	
Chemical Resistance	Low	Medium	High	
Food Compatibility	Low	Medium	High	
Ultraviolet Stability	Low	Medium	High	
Moisture Resistance	Low	Medium	High	
Lubricity	Low	Medium	High	
Electrical Insulation	Low	Medium	High	
Heat Resistance	Low	Medium	High	
Recyclability	Low	Medium	High	
Degradability	Low	Medium	High	
Cost	Low	Medium	High	
Color	Black			
	Gray			
	White			
	Blue			
	Green			
	Red			
	Yellow			

Worksheet 6c: Matching Materials

List up to three materials that best satisfy all the requirements listed in Worksheet 3.

Part C		
1.	2.	3.

Worksheet 7c: List of Matching Materials

SELECTING PROCESSES

Step 8: Define the Form Requirements
Study the problem description, define the form characteristics desired of the product, and list these below in Worksheet 8: List of Form Requirements.

Part A
Part B
Part C

Worksheet 8: List of Form Requirements

Step 9: Determine the Stages by which the Desired Form is Achieved
Determine how the form requirements, described in Worksheet 8: List of Form Requirements, can be achieved as a series of successive stages. List each part and the necessary operations required to achieve the desired form in stages. If there are multiple parts in the product, call them out as well. List this information below in Worksheet 9: Stages of Form Generation.

Stage	What form is achieved in this stage?
	Part A
	Part B
	Part C

Worksheet 9: Stages of Form Generation

Step 10: Determine the Manufacturing Processes that are Relevant
Refer to the materials selections listed previously in Worksheet 7: List of Matching Materials, and the form requirements listed in Worksheet 9: Stages of Form Generation. Apply the procedure described below to determine the processes that are appropriate for each stage of processing each of the selected materials.
- Load MaterialTool and on the Title page click on the button called **"Processes."** This takes you to the Processes section of MaterialTool.
- Click on any of the process groups buttons: **"Finishing"**, **"Forming"**, **"Joining"** or **"Machining."**
  This takes you to the selected processes group page. Notice that all major material families (such as Ceramic, Leather, and Wood) are listed in the definition. By clicking on these words, it is possible to determine the processes applicable to each material group.
- Click on the button with the name of an individual process. This takes you to that particular process page. By reading the process definition, design tips, and visual consequences, determine if the process can be used to achieve the desired form in the particular material.
- For a deeper understanding of the process, click on the green button named "**Show principle**."
  List the selected processes below in Worksheet 10: List of Selected Manufacturing Processes.

Step	Process	Description
Part A		
1.		
2.		
3.		
4.		
5.		
Part B		
1.		
2.		
3.		
4.		
5.		
Part C		
1.		
2.		
3.		
4.		
5.		

Worksheet 10: List of Selected Manufacturing Processes

FINAL SELECTION OF MATERIALS & PROCESSES

Step 11: List the Final Selections of Materials and Processes

Refer to the materials selections listed previously in Worksheet 7: List of Matching Materials, and the form requirements listed in Worksheet 10: List of Selected Manufacturing Processes. List the final choices of materials and manufacturing processes below in the Final Report.

FINAL REPORT	
Part A	
Material Selected	
Properties	
Rationale	
Processes Selected	
Operation	
Rationale	
Part B	
Material Selected	
Properties	
Rationale	
Processes Selected	
Operation	
Rationale	
Part C	
Material Selected	
Properties	
Rationale	
Processes Selected	
Operation	
Rationale	

Worksheet 11: Final Report

Unit 5: Storage Box

Problem
You were asked by a manufacturer of domestic storage containers to design a strong storage box that can be used for storing odds and ends. Since it might contain small valuable domestic articles, such as photographs or even some durable dry foods, it is important that the box not soil the contents. The picture shows the storage box that you have designed. It has two components: a sturdy box and a lid.
The box measures 9" by 9" by 6". What materials and processes were used for manufacturing the box and the lid?

SELECTING MATERIALS

Step 1: Identify the Parts of the Product
The product shown above may be made up of more than one part. Examine the photograph of the product, and identify all the parts of the product. List these in Worksheet 1: List of all Product Parts.

1.	
2.	
3.	
4.	
5.	
6.	
7.	
8.	
9.	
10.	

Worksheet 1: List of all Product Parts

Step 2: List Unique Parts
Some of the parts listed above may be identical parts. Identify the unique parts of the product. Consider mirrored parts, such as a left half and a right half of a product, to be identical parts. List up to three unique parts in Worksheet 2: Unique Product Parts.

Part A	
Part B	
Part C	

Worksheet 2: Unique Product Parts

Step 3: Define the Desired Property Values

First identify what performance characteristics are required for each part that has been identified above. Circle the property values for the material that satisfy the performance requirements for each part in Worksheet 3: Property Values below. Note that only some of the properties will be relevant for each part; leave the rest blank.

Part A							
Hardness		Low		Medium		High	
Impact Strength		Low		Medium		High	
Rigidity		Low		Medium		High	
Abrasion Resistance		Low		Medium		High	
Optical Clarity		Low		Medium		High	
Chemical Resistance		Low		Medium		High	
Food Compatibility		Low		Medium		High	
Ultraviolet Stability		Low		Medium		High	
Moisture Resistance		Low		Medium		High	
Lubricity		Low		Medium		High	
Electrical Insulation		Low		Medium		High	
Heat Resistance		Low		Medium		High	
Recyclability		Low		Medium		High	
Degradability		Low		Medium		High	
Cost		Low		Medium		High	
Color	Black	Gray	White	Blue	Green	Red	Yellow
Part B							
Hardness		Low		Medium		High	
Impact Strength		Low		Medium		High	
Rigidity		Low		Medium		High	
Abrasion Resistance		Low		Medium		High	
Optical Clarity		Low		Medium		High	
Chemical Resistance		Low		Medium		High	
Food Compatibility		Low		Medium		High	
Ultraviolet Stability		Low		Medium		High	
Moisture Resistance		Low		Medium		High	
Lubricity		Low		Medium		High	
Electrical Insulation		Low		Medium		High	
Heat Resistance		Low		Medium		High	
Recyclability		Low		Medium		High	
Degradability		Low		Medium		High	
Cost		Low		Medium		High	
Color	Black	Gray	White	Blue	Green	Red	Yellow
Part C							
Hardness		Low		Medium		High	
Impact Strength		Low		Medium		High	
Rigidity		Low		Medium		High	
Abrasion Resistance		Low		Medium		High	
Optical Clarity		Low		Medium		High	
Chemical Resistance		Low		Medium		High	
Food Compatibility		Low		Medium		High	
Ultraviolet Stability		Low		Medium		High	
Moisture Resistance		Low		Medium		High	
Lubricity		Low		Medium		High	
Electrical Insulation		Low		Medium		High	
Heat Resistance		Low		Medium		High	
Recyclability		Low		Medium		High	
Degradability		Low		Medium		High	
Cost		Low		Medium		High	
Color	Black	Gray	White	Blue	Green	Red	Yellow

Worksheet 3: Property Values

Step 4: Determine Matching Material Families

On the MaterialTool Title page click on the button called "Materials." This takes you to the Materials Section. Click on the large button called "Properties." This takes you to the Materials Properties page. Notice the list of material families listed on the left: Ceramic, Cement, Fiber, and so on. In the right pane you see a list of properties such as Hardness, Impact Strength, Rigidity, and so on. Click on the values Low, Medium, or High to determine Material Families that match the choices from Worksheet 3: Property Values. List the matching families for Part A in Worksheet 4a, the matching families for Part B in Worksheet 4b, and the matching families for Part C in Worksheet 4c.

Part A				
Properties	**Value Selected**			**Matching Materials**
Hardness	Low	Medium	High	
Impact Strength	Low	Medium	High	
Rigidity	Low	Medium	High	
Abrasion Resistance	Low	Medium	High	
Optical Clarity	Low	Medium	High	
Chemical Resistance	Low	Medium	High	
Food Compatibility	Low	Medium	High	
Ultraviolet Stability	Low	Medium	High	
Moisture Resistance	Low	Medium	High	
Lubricity	Low	Medium	High	
Electrical Insulation	Low	Medium	High	
Heat Resistance	Low	Medium	High	
Recyclability	Low	Medium	High	
Degradability	Low	Medium	High	
Cost	Low	Medium	High	
Color	Black			
	Gray			
	White			
	Blue			
	Green			
	Red			
	Yellow			

Worksheet 4a: Matching Material Families

Step 5: Identify the Most Appropriate Material Families

List up to three material families that satisfy all the requirements listed in Worksheet 4a.

Part A					
1.		2.		3.	

Worksheet 5a: List of Matching Material Families

Part B				
Properties	**Value Selected**			**Matching Materials**
Hardness	Low	Medium	High	
Impact Strength	Low	Medium	High	
Rigidity	Low	Medium	High	
Abrasion Resistance	Low	Medium	High	
Optical Clarity	Low	Medium	High	
Chemical Resistance	Low	Medium	High	
Food Compatibility	Low	Medium	High	
Ultraviolet Stability	Low	Medium	High	
Moisture Resistance	Low	Medium	High	
Lubricity	Low	Medium	High	
Electrical Insulation	Low	Medium	High	
Heat Resistance	Low	Medium	High	
Recyclability	Low	Medium	High	
Degradability	Low	Medium	High	
Cost	Low	Medium	High	
Color	Black			
	Gray			
	White			
	Blue			
	Green			
	Red			
	Yellow			

Worksheet 4b: Matching Material Families

List up to three material families that satisfy all the requirements listed in Worksheet 4b.

Part B					
1.		2.		3.	

Worksheet 5b: List of Matching Material Families

Part C				
Properties	**Value Selected**			**Matching Materials**
Hardness	Low	Medium	High	
Impact Strength	Low	Medium	High	
Rigidity	Low	Medium	High	
Abrasion Resistance	Low	Medium	High	
Optical Clarity	Low	Medium	High	
Chemical Resistance	Low	Medium	High	
Food Compatibility	Low	Medium	High	
Ultraviolet Stability	Low	Medium	High	
Moisture Resistance	Low	Medium	High	
Lubricity	Low	Medium	High	
Electrical Insulation	Low	Medium	High	
Heat Resistance	Low	Medium	High	
Recyclability	Low	Medium	High	
Degradability	Low	Medium	High	
Cost	Low	Medium	High	
Color		Black		
		Gray		
		White		
		Blue		
		Green		
		Red		
		Yellow		

Worksheet 4c: Matching Material Families

List up to three material families that satisfy all the requirements listed in Worksheet 4c.

Part C					
1.		2.		3.	

Worksheet 5c: List of Matching Material Families

Step 6: Determine Matching Materials

Once again, go to the Materials Section and click on the large button called "Properties." This takes you to the main Materials Properties page. Referring to the list you created in Worksheet 3, click the arrow next to the first family. This brings up detailed materials. In the right pane, you see a list of properties such as Hardness, Impact Strength, Rigidity, and so on. Once again select the materials that match the choices in Worksheet 1. List these below in Worksheet 6: Matching Materials. List the matching materials for Part A in Worksheet 6a, the matching materials for Part B in Worksheet 6b, and the matching materials for Part C in Worksheet 6c.

Part A				
Properties	**Value Selected**			**Matching Materials**
Hardness	Low	Medium	High	
Impact Strength	Low	Medium	High	
Rigidity	Low	Medium	High	
Abrasion Resistance	Low	Medium	High	
Optical Clarity	Low	Medium	High	
Chemical Resistance	Low	Medium	High	
Food Compatibility	Low	Medium	High	
Ultraviolet Stability	Low	Medium	High	
Moisture Resistance	Low	Medium	High	
Lubricity	Low	Medium	High	
Electrical Insulation	Low	Medium	High	
Heat Resistance	Low	Medium	High	
Recyclability	Low	Medium	High	
Degradability	Low	Medium	High	
Cost	Low	Medium	High	
Color	Black			
	Gray			
	White			
	Blue			
	Green			
	Red			
	Yellow			

Worksheet 6a: Matching Materials

List up to three materials that best satisfy all the requirements listed in Worksheet 3.

Part A		
1.	2.	3.

Worksheet 7a: List of Matching Materials

Part B				
Properties	Value Selected			Matching Materials
Hardness	Low	Medium	High	
Impact Strength	Low	Medium	High	
Rigidity	Low	Medium	High	
Abrasion Resistance	Low	Medium	High	
Optical Clarity	Low	Medium	High	
Chemical Resistance	Low	Medium	High	
Food Compatibility	Low	Medium	High	
Ultraviolet Stability	Low	Medium	High	
Moisture Resistance	Low	Medium	High	
Lubricity	Low	Medium	High	
Electrical Insulation	Low	Medium	High	
Heat Resistance	Low	Medium	High	
Recyclability	Low	Medium	High	
Degradability	Low	Medium	High	
Cost	Low	Medium	High	
Color	Black			
	Gray			
	White			
	Blue			
	Green			
	Red			
	Yellow			

Worksheet 6b: Matching Materials

List up to three materials that best satisfy all the requirements listed in Worksheet 3.

Part B					
1.		2.		3.	

Worksheet 7b: List of Matching Materials

Part C				
Properties	**Value Selected**			**Matching Materials**
Hardness	Low	Medium	High	
Impact Strength	Low	Medium	High	
Rigidity	Low	Medium	High	
Abrasion Resistance	Low	Medium	High	
Optical Clarity	Low	Medium	High	
Chemical Resistance	Low	Medium	High	
Food Compatibility	Low	Medium	High	
Ultraviolet Stability	Low	Medium	High	
Moisture Resistance	Low	Medium	High	
Lubricity	Low	Medium	High	
Electrical Insulation	Low	Medium	High	
Heat Resistance	Low	Medium	High	
Recyclability	Low	Medium	High	
Degradability	Low	Medium	High	
Cost	Low	Medium	High	
Color	Black			
	Gray			
	White			
	Blue			
	Green			
	Red			
	Yellow			

Worksheet 6c: Matching Materials

List up to three materials that best satisfy all the requirements listed in Worksheet 3.

Part C		
1.	2.	3.

Worksheet 7c: List of Matching Materials

SELECTING PROCESSES

Step 8: Define the Form Requirements
Study the problem description, define the form characteristics desired of the product, and list these below in Worksheet 8: List of Form Requirements.

Part A
Part B
Part C

Worksheet 8: List of Form Requirements

Step 9: Determine the Stages by which the Desired Form is Achieved
Determine how the form requirements, described in Worksheet 8: List of Form Requirements, can be achieved as a series of successive stages. List each part and the necessary operations required to achieve the desired form in stages. If there are multiple parts in the product, call them out as well. List this information below in Worksheet 9: Stages of Form Generation.

Stage	What form is achieved in this stage?
	Part A
	Part B
	Part C

Worksheet 9: Stages of Form Generation

Step 10: Determine the Manufacturing Processes that are Relevant
Refer to the materials selections listed previously in Worksheet 7: List of Matching Materials, and the form requirements listed in Worksheet 9: Stages of Form Generation. Apply the procedure described below to determine the processes that are appropriate for each stage of processing each of the selected materials.

- Load MaterialTool and on the Title page click on the button called "**Processes.**" This takes you to the Processes section of MaterialTool.
- Click on any of the process groups buttons: "**Finishing**", "**Forming**", "**Joining**" or "**Machining**."
  This takes you to the selected processes group page. Notice that all major material families (such as Ceramic, Leather, and Wood) are listed in the definition. By clicking on these words, it is possible to determine the processes applicable to each material group.
- Click on the button with the name of an individual process. This takes you to that particular process page. By reading the process definition, design tips, and visual consequences, determine if the process can be used to achieve the desired form in the particular material.
- For a deeper understanding of the process, click on the green button named "**Show principle**."
  List the selected processes below in Worksheet 10: List of Selected Manufacturing Processes.

Step	Process	Description
Part A		
1.		
2.		
3.		
4.		
5.		
Part B		
1.		
2.		
3.		
4.		
5.		
Part C		
1.		
2.		
3.		
4.		
5.		

Worksheet 10: List of Selected Manufacturing Processes

FINAL SELECTION OF MATERIALS & PROCESSES

Step 11: List the Final Selections of Materials and Processes
Refer to the materials selections listed previously in Worksheet 7: List of Matching Materials, and the form requirements listed in Worksheet 10: List of Selected Manufacturing Processes. List the final choices of materials and manufacturing processes below in the Final Report.

FINAL REPORT	
Part A	
Material Selected	
Properties	
Rationale	
Processes Selected	
Operation	
Rationale	
Part B	
Material Selected	
Properties	
Rationale	
Processes Selected	
Operation	
Rationale	
Part C	
Material Selected	
Properties	
Rationale	
Processes Selected	
Operation	
Rationale	

Worksheet 11: Final Report

Unit 6: Screen Construction Block

Problem
You were asked by a manufacturer of construction materials to develop a block that could be used for constructing outdoor screens. The main requirement is that the blocks be produceable very inexpensively. The picture shows the building block that you designed. Describe the materials you selected for this product and the processes that you specified for its manufacture.

SELECTING MATERIALS

Step 1: Identify the Parts of the Product
The product shown above may be made up of more than one part. Examine the photograph of the product, and identify all the parts of the product. List these in Worksheet 1: List of all Product Parts.

1.	
2.	
3.	
4.	
5.	
6.	
7.	
8.	
9.	
10.	

Worksheet 1: List of all Product Parts

Step 2: List Unique Parts
Some of the parts listed above may be identical parts. Identify the unique parts of the product. Consider mirrored parts, such as a left half and a right half of a product, to be identical parts. List up to three unique parts in Worksheet 2: Unique Product Parts.

Part A	
Part B	
Part C	

Worksheet 2: Unique Product Parts

Step 3: Define the Desired Property Values

First identify what performance characteristics are required for each part that has been identified above. Circle the property values for the material that satisfy the performance requirements for each part in Worksheet 3: Property Values below. Note that only some of the properties will be relevant for each part; leave the rest blank.

Part A							
Hardness	Low		Medium			High	
Impact Strength	Low		Medium			High	
Rigidity	Low		Medium			High	
Abrasion Resistance	Low		Medium			High	
Optical Clarity	Low		Medium			High	
Chemical Resistance	Low		Medium			High	
Food Compatibility	Low		Medium			High	
Ultraviolet Stability	Low		Medium			High	
Moisture Resistance	Low		Medium			High	
Lubricity	Low		Medium			High	
Electrical Insulation	Low		Medium			High	
Heat Resistance	Low		Medium			High	
Recyclability	Low		Medium			High	
Degradability	Low		Medium			High	
Cost	Low		Medium			High	
Color	Black	Gray	White	Blue	Green	Red	Yellow
Part B							
Hardness	Low		Medium			High	
Impact Strength	Low		Medium			High	
Rigidity	Low		Medium			High	
Abrasion Resistance	Low		Medium			High	
Optical Clarity	Low		Medium			High	
Chemical Resistance	Low		Medium			High	
Food Compatibility	Low		Medium			High	
Ultraviolet Stability	Low		Medium			High	
Moisture Resistance	Low		Medium			High	
Lubricity	Low		Medium			High	
Electrical Insulation	Low		Medium			High	
Heat Resistance	Low		Medium			High	
Recyclability	Low		Medium			High	
Degradability	Low		Medium			High	
Cost	Low		Medium			High	
Color	Black	Gray	White	Blue	Green	Red	Yellow
Part C							
Hardness	Low		Medium			High	
Impact Strength	Low		Medium			High	
Rigidity	Low		Medium			High	
Abrasion Resistance	Low		Medium			High	
Optical Clarity	Low		Medium			High	
Chemical Resistance	Low		Medium			High	
Food Compatibility	Low		Medium			High	
Ultraviolet Stability	Low		Medium			High	
Moisture Resistance	Low		Medium			High	
Lubricity	Low		Medium			High	
Electrical Insulation	Low		Medium			High	
Heat Resistance	Low		Medium			High	
Recyclability	Low		Medium			High	
Degradability	Low		Medium			High	
Cost	Low		Medium			High	
Color	Black	Gray	White	Blue	Green	Red	Yellow

Worksheet 3: Property Values

Step 4: Determine Matching Material Families

On the MaterialTool Title page click on the button called "Materials." This takes you to the Materials Section. Click on the large button called "Properties." This takes you to the Materials Properties page. Notice the list of material families listed on the left: Ceramic, Cement, Fiber, and so on. In the right pane you see a list of properties such as Hardness, Impact Strength, Rigidity, and so on. Click on the values Low, Medium, or High to determine Material Families that match the choices from Worksheet 3: Property Values. List the matching families for Part A in Worksheet 4a, the matching families for Part B in Worksheet 4b, and the matching families for Part C in Worksheet 4c.

Part A				
Properties	**Value Selected**			**Matching Materials**
Hardness	Low	Medium	High	
Impact Strength	Low	Medium	High	
Rigidity	Low	Medium	High	
Abrasion Resistance	Low	Medium	High	
Optical Clarity	Low	Medium	High	
Chemical Resistance	Low	Medium	High	
Food Compatibility	Low	Medium	High	
Ultraviolet Stability	Low	Medium	High	
Moisture Resistance	Low	Medium	High	
Lubricity	Low	Medium	High	
Electrical Insulation	Low	Medium	High	
Heat Resistance	Low	Medium	High	
Recyclability	Low	Medium	High	
Degradability	Low	Medium	High	
Cost	Low	Medium	High	
Color	Black			
	Gray			
	White			
	Blue			
	Green			
	Red			
	Yellow			

Worksheet 4a: Matching Material Families

Step 5: Identify the Most Appropriate Material Families

List up to three material families that satisfy all the requirements listed in Worksheet 4a.

Part A					
1.		2.		3.	

Worksheet 5a: List of Matching Material Families

Part B				
Properties	**Value Selected**			**Matching Materials**
Hardness	Low	Medium	High	
Impact Strength	Low	Medium	High	
Rigidity	Low	Medium	High	
Abrasion Resistance	Low	Medium	High	
Optical Clarity	Low	Medium	High	
Chemical Resistance	Low	Medium	High	
Food Compatibility	Low	Medium	High	
Ultraviolet Stability	Low	Medium	High	
Moisture Resistance	Low	Medium	High	
Lubricity	Low	Medium	High	
Electrical Insulation	Low	Medium	High	
Heat Resistance	Low	Medium	High	
Recyclability	Low	Medium	High	
Degradability	Low	Medium	High	
Cost	Low	Medium	High	
Color	Black			
	Gray			
	White			
	Blue			
	Green			
	Red			
	Yellow			

Worksheet 4b: Matching Material Families

List up to three material families that satisfy all the requirements listed in Worksheet 4b.

Part B		
1.	2.	3.

Worksheet 5b: List of Matching Material Families

Part C				
Properties	**Value Selected**			**Matching Materials**
Hardness	Low	Medium	High	
Impact Strength	Low	Medium	High	
Rigidity	Low	Medium	High	
Abrasion Resistance	Low	Medium	High	
Optical Clarity	Low	Medium	High	
Chemical Resistance	Low	Medium	High	
Food Compatibility	Low	Medium	High	
Ultraviolet Stability	Low	Medium	High	
Moisture Resistance	Low	Medium	High	
Lubricity	Low	Medium	High	
Electrical Insulation	Low	Medium	High	
Heat Resistance	Low	Medium	High	
Recyclability	Low	Medium	High	
Degradability	Low	Medium	High	
Cost	Low	Medium	High	
Color	Black			
	Gray			
	White			
	Blue			
	Green			
	Red			
	Yellow			

Worksheet 4c: Matching Material Families

List up to three material families that satisfy all the requirements listed in Worksheet 4c.

Part C		
1.	2.	3.

Worksheet 5c: List of Matching Material Families

Step 6: Determine Matching Materials
Once again, go to the Materials Section and click on the large button called "Properties." This takes you to the main Materials Properties page. Referring to the list you created in Worksheet 3, click the arrow next to the first family. This brings up detailed materials. In the right pane, you see a list of properties such as Hardness, Impact Strength, Rigidity, and so on. Once again select the materials that match the choices in Worksheet 1. List these below in Worksheet 6: Matching Materials. List the matching materials for Part A in Worksheet 6a, the matching materials for Part B in Worksheet 6b, and the matching materials for Part C in Worksheet 6c.

Part A				
Properties	**Value Selected**			**Matching Materials**
Hardness	Low	Medium	High	
Impact Strength	Low	Medium	High	
Rigidity	Low	Medium	High	
Abrasion Resistance	Low	Medium	High	
Optical Clarity	Low	Medium	High	
Chemical Resistance	Low	Medium	High	
Food Compatibility	Low	Medium	High	
Ultraviolet Stability	Low	Medium	High	
Moisture Resistance	Low	Medium	High	
Lubricity	Low	Medium	High	
Electrical Insulation	Low	Medium	High	
Heat Resistance	Low	Medium	High	
Recyclability	Low	Medium	High	
Degradability	Low	Medium	High	
Cost	Low	Medium	High	
Color	Black			
	Gray			
	White			
	Blue			
	Green			
	Red			
	Yellow			

Worksheet 6a: Matching Materials

List up to three materials that best satisfy all the requirements listed in Worksheet 3.

Part A		
1.	2.	3.

Worksheet 7a: List of Matching Materials

Part B				
Properties	**Value Selected**			**Matching Materials**
Hardness	Low	Medium	High	
Impact Strength	Low	Medium	High	
Rigidity	Low	Medium	High	
Abrasion Resistance	Low	Medium	High	
Optical Clarity	Low	Medium	High	
Chemical Resistance	Low	Medium	High	
Food Compatibility	Low	Medium	High	
Ultraviolet Stability	Low	Medium	High	
Moisture Resistance	Low	Medium	High	
Lubricity	Low	Medium	High	
Electrical Insulation	Low	Medium	High	
Heat Resistance	Low	Medium	High	
Recyclability	Low	Medium	High	
Degradability	Low	Medium	High	
Cost	Low	Medium	High	
Color		Black		
		Gray		
		White		
		Blue		
		Green		
		Red		
		Yellow		

Worksheet 6b: Matching Materials

List up to three materials that best satisfy all the requirements listed in Worksheet 3.

Part B					
1.		2.		3.	

Worksheet 7b: List of Matching Materials

Part C				
Properties	**Value Selected**			**Matching Materials**
Hardness	Low	Medium	High	
Impact Strength	Low	Medium	High	
Rigidity	Low	Medium	High	
Abrasion Resistance	Low	Medium	High	
Optical Clarity	Low	Medium	High	
Chemical Resistance	Low	Medium	High	
Food Compatibility	Low	Medium	High	
Ultraviolet Stability	Low	Medium	High	
Moisture Resistance	Low	Medium	High	
Lubricity	Low	Medium	High	
Electrical Insulation	Low	Medium	High	
Heat Resistance	Low	Medium	High	
Recyclability	Low	Medium	High	
Degradability	Low	Medium	High	
Cost	Low	Medium	High	
Color	Black			
	Gray			
	White			
	Blue			
	Green			
	Red			
	Yellow			

Worksheet 6c: Matching Materials

List up to three materials that best satisfy all the requirements listed in Worksheet 3.

Part C		
1.	2.	3.

Worksheet 7c: List of Matching Materials

SELECTING PROCESSES

Step 8: Define the Form Requirements
Study the problem description, define the form characteristics desired of the product, and list these below in Worksheet 8: List of Form Requirements.

Part A
Part B
Part C

Worksheet 8: List of Form Requirements

Step 9: Determine the Stages by which the Desired Form is Achieved
Determine how the form requirements, described in Worksheet 8: List of Form Requirements, can be achieved as a series of successive stages. List each part and the necessary operations required to achieve the desired form in stages. If there are multiple parts in the product, call them out as well. List this information below in Worksheet 9: Stages of Form Generation.

Stage	What form is achieved in this stage?
Part A	
Part B	
Part C	

Worksheet 9: Stages of Form Generation

Step 10: Determine the Manufacturing Processes that are Relevant
Refer to the materials selections listed previously in Worksheet 7: List of Matching Materials, and the form requirements listed in Worksheet 9: Stages of Form Generation. Apply the procedure described below to determine the processes that are appropriate for each stage of processing each of the selected materials.
- Load MaterialTool and on the Title page click on the button called "**Processes.**" This takes you to the Processes section of MaterialTool.
- Click on any of the process groups buttons: "**Finishing**", "**Forming**", "**Joining**" or "**Machining**."
  This takes you to the selected processes group page. Notice that all major material families (such as Ceramic, Leather, and Wood) are listed in the definition. By clicking on these words, it is possible to determine the processes applicable to each material group.
- Click on the button with the name of an individual process. This takes you to that particular process page. By reading the process definition, design tips, and visual consequences, determine if the process can be used to achieve the desired form in the particular material.
- For a deeper understanding of the process, click on the green button named "**Show principle**."
  List the selected processes below in Worksheet 10: List of Selected Manufacturing Processes.

Step	Process	Description
Part A		
1.		
2.		
3.		
4.		
5.		
Part B		
1.		
2.		
3.		
4.		
5.		
Part C		
1.		
2.		
3.		
4.		
5.		

Worksheet 10: List of Selected Manufacturing Processes

FINAL SELECTION OF MATERIALS & PROCESSES

Step 11: List the Final Selections of Materials and Processes
Refer to the materials selections listed previously in Worksheet 7: List of Matching Materials, and the form requirements listed in Worksheet 10: List of Selected Manufacturing Processes. List the final choices of materials and manufacturing processes below in the Final Report.

\multicolumn{2}{c}{FINAL REPORT}
Part A	
Material Selected	
Properties	
Rationale	
Processes Selected	
Operation	
Rationale	
Part B	
Material Selected	
Properties	
Rationale	
Processes Selected	
Operation	
Rationale	
Part C	
Material Selected	
Properties	
Rationale	
Processes Selected	
Operation	
Rationale	

Worksheet 11: Final Report

Unit 7: Teapot

Problem
The picture shows a teapot, which is used for brewing tea by pouring boiling water into it. It is available at a reasonable price, and consists of two parts: the pot itself and a lid. Thousands of these teapots are manufactured each month. What materials are each of the two parts of the teapot made of?
What are the processes used for manufacturing them?

SELECTING MATERIALS

Step 1: Identify the Parts of the Product
The product shown above may be made up of more than one part. Examine the photograph of the product, and identify all the parts of the product. List these in Worksheet 1: List of all Product Parts.

1.	
2.	
3.	
4.	
5.	
6.	
7.	
8.	
9.	
10.	

Worksheet 1: List of all Product Parts

Step 2: List Unique Parts
Some of the parts listed above may be identical parts. Identify the unique parts of the product. Consider mirrored parts, such as a left half and a right half of a product, to be identical parts. List up to three unique parts in Worksheet 2: Unique Product Parts.

Part A	
Part B	
Part C	

Worksheet 2: Unique Product Parts

Step 3: Define the Desired Property Values

First identify what performance characteristics are required for each part that has been identified above. Circle the property values for the material that satisfy the performance requirements for each part in Worksheet 3: Property Values below. Note that only some of the properties will be relevant for each part; leave the rest blank.

Part A

Hardness	Low		Medium		High		
Impact Strength	Low		Medium		High		
Rigidity	Low		Medium		High		
Abrasion Resistance	Low		Medium		High		
Optical Clarity	Low		Medium		High		
Chemical Resistance	Low		Medium		High		
Food Compatibility	Low		Medium		High		
Ultraviolet Stability	Low		Medium		High		
Moisture Resistance	Low		Medium		High		
Lubricity	Low		Medium		High		
Electrical Insulation	Low		Medium		High		
Heat Resistance	Low		Medium		High		
Recyclability	Low		Medium		High		
Degradability	Low		Medium		High		
Cost	Low		Medium		High		
Color	Black	Gray	White	Blue	Green	Red	Yellow

Part B

Hardness	Low		Medium		High		
Impact Strength	Low		Medium		High		
Rigidity	Low		Medium		High		
Abrasion Resistance	Low		Medium		High		
Optical Clarity	Low		Medium		High		
Chemical Resistance	Low		Medium		High		
Food Compatibility	Low		Medium		High		
Ultraviolet Stability	Low		Medium		High		
Moisture Resistance	Low		Medium		High		
Lubricity	Low		Medium		High		
Electrical Insulation	Low		Medium		High		
Heat Resistance	Low		Medium		High		
Recyclability	Low		Medium		High		
Degradability	Low		Medium		High		
Cost	Low		Medium		High		
Color	Black	Gray	White	Blue	Green	Red	Yellow

Part C

Hardness	Low		Medium		High		
Impact Strength	Low		Medium		High		
Rigidity	Low		Medium		High		
Abrasion Resistance	Low		Medium		High		
Optical Clarity	Low		Medium		High		
Chemical Resistance	Low		Medium		High		
Food Compatibility	Low		Medium		High		
Ultraviolet Stability	Low		Medium		High		
Moisture Resistance	Low		Medium		High		
Lubricity	Low		Medium		High		
Electrical Insulation	Low		Medium		High		
Heat Resistance	Low		Medium		High		
Recyclability	Low		Medium		High		
Degradability	Low		Medium		High		
Cost	Low		Medium		High		
Color	Black	Gray	White	Blue	Green	Red	Yellow

Worksheet 3: Property Values

Step 4: Determine Matching Material Families

On the MaterialTool Title page click on the button called "Materials." This takes you to the Materials Section. Click on the large button called "Properties." This takes you to the Materials Properties page. Notice the list of material families listed on the left: Ceramic, Cement, Fiber, and so on. In the right pane you see a list of properties such as Hardness, Impact Strength, Rigidity, and so on. Click on the values Low, Medium, or High to determine Material Families that match the choices from Worksheet 3: Property Values. List the matching families for Part A in Worksheet 4a, the matching families for Part B in Worksheet 4b, and the matching families for Part C in Worksheet 4c.

Part A				
Properties	**Value Selected**			**Matching Materials**
Hardness	Low	Medium	High	
Impact Strength	Low	Medium	High	
Rigidity	Low	Medium	High	
Abrasion Resistance	Low	Medium	High	
Optical Clarity	Low	Medium	High	
Chemical Resistance	Low	Medium	High	
Food Compatibility	Low	Medium	High	
Ultraviolet Stability	Low	Medium	High	
Moisture Resistance	Low	Medium	High	
Lubricity	Low	Medium	High	
Electrical Insulation	Low	Medium	High	
Heat Resistance	Low	Medium	High	
Recyclability	Low	Medium	High	
Degradability	Low	Medium	High	
Cost	Low	Medium	High	
Color	Black			
	Gray			
	White			
	Blue			
	Green			
	Red			
	Yellow			

Worksheet 4a: Matching Material Families

Step 5: Identify the Most Appropriate Material Families

List up to three material families that satisfy all the requirements listed in Worksheet 4a.

Part A					
1.		2.		3.	

Worksheet 5a: List of Matching Material Families

Part B				
Properties	**Value Selected**			**Matching Materials**
Hardness	Low	Medium	High	
Impact Strength	Low	Medium	High	
Rigidity	Low	Medium	High	
Abrasion Resistance	Low	Medium	High	
Optical Clarity	Low	Medium	High	
Chemical Resistance	Low	Medium	High	
Food Compatibility	Low	Medium	High	
Ultraviolet Stability	Low	Medium	High	
Moisture Resistance	Low	Medium	High	
Lubricity	Low	Medium	High	
Electrical Insulation	Low	Medium	High	
Heat Resistance	Low	Medium	High	
Recyclability	Low	Medium	High	
Degradability	Low	Medium	High	
Cost	Low	Medium	High	
Color	Black			
	Gray			
	White			
	Blue			
	Green			
	Red			
	Yellow			

Worksheet 4b: Matching Material Families

List up to three material families that satisfy all the requirements listed in Worksheet 4b.

Part B					
1.		2.		3.	

Worksheet 5b: List of Matching Material Families

Part C				
Properties	**Value Selected**			**Matching Materials**
Hardness	Low	Medium	High	
Impact Strength	Low	Medium	High	
Rigidity	Low	Medium	High	
Abrasion Resistance	Low	Medium	High	
Optical Clarity	Low	Medium	High	
Chemical Resistance	Low	Medium	High	
Food Compatibility	Low	Medium	High	
Ultraviolet Stability	Low	Medium	High	
Moisture Resistance	Low	Medium	High	
Lubricity	Low	Medium	High	
Electrical Insulation	Low	Medium	High	
Heat Resistance	Low	Medium	High	
Recyclability	Low	Medium	High	
Degradability	Low	Medium	High	
Cost	Low	Medium	High	
Color	Black			
	Gray			
	White			
	Blue			
	Green			
	Red			
	Yellow			

Worksheet 4c: Matching Material Families

List up to three material families that satisfy all the requirements listed in Worksheet 4c.

Part C		
1.	2.	3.

Worksheet 5c: List of Matching Material Families

Step 6: Determine Matching Materials
Once again, go to the Materials Section and click on the large button called "Properties." This takes you to the main Materials Properties page. Referring to the list you created in Worksheet 3, click the arrow next to the first family. This brings up detailed materials. In the right pane, you see a list of properties such as Hardness, Impact Strength, Rigidity, and so on. Once again select the materials that match the choices in Worksheet 1. List these below in Worksheet 6: Matching Materials. List the matching materials for Part A in Worksheet 6a, the matching materials for Part B in Worksheet 6b, and the matching materials for Part C in Worksheet 6c.

Part A				
Properties	**Value Selected**			**Matching Materials**
Hardness	Low	Medium	High	
Impact Strength	Low	Medium	High	
Rigidity	Low	Medium	High	
Abrasion Resistance	Low	Medium	High	
Optical Clarity	Low	Medium	High	
Chemical Resistance	Low	Medium	High	
Food Compatibility	Low	Medium	High	
Ultraviolet Stability	Low	Medium	High	
Moisture Resistance	Low	Medium	High	
Lubricity	Low	Medium	High	
Electrical Insulation	Low	Medium	High	
Heat Resistance	Low	Medium	High	
Recyclability	Low	Medium	High	
Degradability	Low	Medium	High	
Cost	Low	Medium	High	
Color	Black			
	Gray			
	White			
	Blue			
	Green			
	Red			
	Yellow			

Worksheet 6a: Matching Materials

List up to three materials that best satisfy all the requirements listed in Worksheet 3.

Part A		
1.	2.	3.

Worksheet 7a: List of Matching Materials

Part B				
Properties	**Value Selected**			**Matching Materials**
Hardness	Low	Medium	High	
Impact Strength	Low	Medium	High	
Rigidity	Low	Medium	High	
Abrasion Resistance	Low	Medium	High	
Optical Clarity	Low	Medium	High	
Chemical Resistance	Low	Medium	High	
Food Compatibility	Low	Medium	High	
Ultraviolet Stability	Low	Medium	High	
Moisture Resistance	Low	Medium	High	
Lubricity	Low	Medium	High	
Electrical Insulation	Low	Medium	High	
Heat Resistance	Low	Medium	High	
Recyclability	Low	Medium	High	
Degradability	Low	Medium	High	
Cost	Low	Medium	High	
Color	Black			
	Gray			
	White			
	Blue			
	Green			
	Red			
	Yellow			

Worksheet 6b: Matching Materials

List up to three materials that best satisfy all the requirements listed in Worksheet 3.

Part B					
1.		2.		3.	

Worksheet 7b: List of Matching Materials

Part C				
Properties	**Value Selected**			**Matching Materials**
Hardness	Low	Medium	High	
Impact Strength	Low	Medium	High	
Rigidity	Low	Medium	High	
Abrasion Resistance	Low	Medium	High	
Optical Clarity	Low	Medium	High	
Chemical Resistance	Low	Medium	High	
Food Compatibility	Low	Medium	High	
Ultraviolet Stability	Low	Medium	High	
Moisture Resistance	Low	Medium	High	
Lubricity	Low	Medium	High	
Electrical Insulation	Low	Medium	High	
Heat Resistance	Low	Medium	High	
Recyclability	Low	Medium	High	
Degradability	Low	Medium	High	
Cost	Low	Medium	High	
Color	Black			
	Gray			
	White			
	Blue			
	Green			
	Red			
	Yellow			

Worksheet 6c: Matching Materials

List up to three materials that best satisfy all the requirements listed in Worksheet 3.

Part C		
1.	2.	3.

Worksheet 7c: List of Matching Materials

SELECTING PROCESSES

Step 8: Define the Form Requirements
Study the problem description, define the form characteristics desired of the product, and list these below in Worksheet 8: List of Form Requirements.

Part A

Part B

Part C

Worksheet 8: List of Form Requirements

Step 9: Determine the Stages by which the Desired Form is Achieved
Determine how the form requirements, described in Worksheet 8: List of Form Requirements, can be achieved as a series of successive stages. List each part and the necessary operations required to achieve the desired form in stages. If there are multiple parts in the product, call them out as well. List this information below in Worksheet 9: Stages of Form Generation.

Stage	What form is achieved in this stage?
Part A	
Part B	
Part C	

Worksheet 9: Stages of Form Generation

Step 10: Determine the Manufacturing Processes that are Relevant
Refer to the materials selections listed previously in Worksheet 7: List of Matching Materials, and the form requirements listed in Worksheet 9: Stages of Form Generation. Apply the procedure described below to determine the processes that are appropriate for each stage of processing each of the selected materials.
- Load MaterialTool and on the Title page click on the button called "**Processes.**" This takes you to the Processes section of MaterialTool.
- Click on any of the process groups buttons: "**Finishing**", "**Forming**", "**Joining**" or "**Machining**." This takes you to the selected processes group page. Notice that all major material families (such as Ceramic, Leather, and Wood) are listed in the definition. By clicking on these words, it is possible to determine the processes applicable to each material group.
- Click on the button with the name of an individual process. This takes you to that particular process page. By reading the process definition, design tips, and visual consequences, determine if the process can be used to achieve the desired form in the particular material.
- For a deeper understanding of the process, click on the green button named "**Show principle**."
List the selected processes below in Worksheet 10: List of Selected Manufacturing Processes.

Step	Process	Description
Part A		
1.		
2.		
3.		
4.		
5.		
Part B		
1.		
2.		
3.		
4.		
5.		
Part C		
1.		
2.		
3.		
4.		
5.		

Worksheet 10: List of Selected Manufacturing Processes

FINAL SELECTION OF MATERIALS & PROCESSES

Step 11: List the Final Selections of Materials and Processes

Refer to the materials selections listed previously in Worksheet 7: List of Matching Materials, and the form requirements listed in Worksheet 10: List of Selected Manufacturing Processes. List the final choices of materials and manufacturing processes below in the Final Report.

FINAL REPORT	
Part A	
Material Selected	
Properties	
Rationale	
Processes Selected	
Operation	
Rationale	
Part B	
Material Selected	
Properties	
Rationale	
Processes Selected	
Operation	
Rationale	
Part C	
Material Selected	
Properties	
Rationale	
Processes Selected	
Operation	
Rationale	

Worksheet 11: Final Report

Unit 8: Nail Clipper

Problem
The picture shows a common nail clipper that can be used for trimming and filing nails. It must be sharp so that it cuts cleanly through nails and keeps its keen edge. The nail file should also keep its filing action without getting smoothed by repeated use. The nail clipper is often kept in the bathroom where moisture is present, and it is important that water not affect it. Identify the major components of the nail clipper, and the materials and processes used for manufacturing it.

SELECTING MATERIALS

Step 1: Identify the Parts of the Product
The product shown above may be made up of more than one part. Examine the photograph of the product, and identify all the parts of the product. List these in Worksheet 1: List of all Product Parts.

1.	
2.	
3.	
4.	
5.	
6.	
7.	
8.	
9.	
10.	

Worksheet 1: List of all Product Parts

Step 2: List Unique Parts
Some of the parts listed above may be identical parts. Identify the unique parts of the product. Consider mirrored parts, such as a left half and a right half of a product, to be identical parts. List up to three unique parts in Worksheet 2: Unique Product Parts.

Part A	
Part B	
Part C	

Worksheet 2: Unique Product Parts

Step 3: Define the Desired Property Values

First identify what performance characteristics are required for each part that has been identified above. Circle the property values for the material that satisfy the performance requirements for each part in Worksheet 3: Property Values below. Note that only some of the properties will be relevant for each part; leave the rest blank.

Part A							
Hardness		Low		Medium		High	
Impact Strength		Low		Medium		High	
Rigidity		Low		Medium		High	
Abrasion Resistance		Low		Medium		High	
Optical Clarity		Low		Medium		High	
Chemical Resistance		Low		Medium		High	
Food Compatibility		Low		Medium		High	
Ultraviolet Stability		Low		Medium		High	
Moisture Resistance		Low		Medium		High	
Lubricity		Low		Medium		High	
Electrical Insulation		Low		Medium		High	
Heat Resistance		Low		Medium		High	
Recyclability		Low		Medium		High	
Degradability		Low		Medium		High	
Cost		Low		Medium		High	
Color	Black	Gray	White	Blue	Green	Red	Yellow
Part B							
Hardness		Low		Medium		High	
Impact Strength		Low		Medium		High	
Rigidity		Low		Medium		High	
Abrasion Resistance		Low		Medium		High	
Optical Clarity		Low		Medium		High	
Chemical Resistance		Low		Medium		High	
Food Compatibility		Low		Medium		High	
Ultraviolet Stability		Low		Medium		High	
Moisture Resistance		Low		Medium		High	
Lubricity		Low		Medium		High	
Electrical Insulation		Low		Medium		High	
Heat Resistance		Low		Medium		High	
Recyclability		Low		Medium		High	
Degradability		Low		Medium		High	
Cost		Low		Medium		High	
Color	Black	Gray	White	Blue	Green	Red	Yellow
Part C							
Hardness		Low		Medium		High	
Impact Strength		Low		Medium		High	
Rigidity		Low		Medium		High	
Abrasion Resistance		Low		Medium		High	
Optical Clarity		Low		Medium		High	
Chemical Resistance		Low		Medium		High	
Food Compatibility		Low		Medium		High	
Ultraviolet Stability		Low		Medium		High	
Moisture Resistance		Low		Medium		High	
Lubricity		Low		Medium		High	
Electrical Insulation		Low		Medium		High	
Heat Resistance		Low		Medium		High	
Recyclability		Low		Medium		High	
Degradability		Low		Medium		High	
Cost		Low		Medium		High	
Color	Black	Gray	White	Blue	Green	Red	Yellow

Worksheet 3: Property Values

Step 4: Determine Matching Material Families

On the MaterialTool Title page click on the button called "Materials." This takes you to the Materials Section. Click on the large button called "Properties." This takes you to the Materials Properties page. Notice the list of material families listed on the left: Ceramic, Cement, Fiber, and so on. In the right pane you see a list of properties such as Hardness, Impact Strength, Rigidity, and so on. Click on the values Low, Medium, or High to determine Material Families that match the choices from Worksheet 3: Property Values. List the matching families for Part A in Worksheet 4a, the matching families for Part B in Worksheet 4b, and the matching families for Part C in Worksheet 4c.

Part A				
Properties	**Value Selected**			**Matching Materials**
Hardness	Low	Medium	High	
Impact Strength	Low	Medium	High	
Rigidity	Low	Medium	High	
Abrasion Resistance	Low	Medium	High	
Optical Clarity	Low	Medium	High	
Chemical Resistance	Low	Medium	High	
Food Compatibility	Low	Medium	High	
Ultraviolet Stability	Low	Medium	High	
Moisture Resistance	Low	Medium	High	
Lubricity	Low	Medium	High	
Electrical Insulation	Low	Medium	High	
Heat Resistance	Low	Medium	High	
Recyclability	Low	Medium	High	
Degradability	Low	Medium	High	
Cost	Low	Medium	High	
Color		Black		
		Gray		
		White		
		Blue		
		Green		
		Red		
		Yellow		

Worksheet 4a: Matching Material Families

Step 5: Identify the Most Appropriate Material Families

List up to three material families that satisfy all the requirements listed in Worksheet 4a.

Part A		
1.	2.	3.

Worksheet 5a: List of Matching Material Families

Part B				
Properties	**Value Selected**			**Matching Materials**
Hardness	Low	Medium	High	
Impact Strength	Low	Medium	High	
Rigidity	Low	Medium	High	
Abrasion Resistance	Low	Medium	High	
Optical Clarity	Low	Medium	High	
Chemical Resistance	Low	Medium	High	
Food Compatibility	Low	Medium	High	
Ultraviolet Stability	Low	Medium	High	
Moisture Resistance	Low	Medium	High	
Lubricity	Low	Medium	High	
Electrical Insulation	Low	Medium	High	
Heat Resistance	Low	Medium	High	
Recyclability	Low	Medium	High	
Degradability	Low	Medium	High	
Cost	Low	Medium	High	
Color	Black			
	Gray			
	White			
	Blue			
	Green			
	Red			
	Yellow			

Worksheet 4b: Matching Material Families

List up to three material families that satisfy all the requirements listed in Worksheet 4b.

Part B					
1.		2.		3.	

Worksheet 5b: List of Matching Material Families

Part C				
Properties	**Value Selected**			**Matching Materials**
Hardness	Low	Medium	High	
Impact Strength	Low	Medium	High	
Rigidity	Low	Medium	High	
Abrasion Resistance	Low	Medium	High	
Optical Clarity	Low	Medium	High	
Chemical Resistance	Low	Medium	High	
Food Compatibility	Low	Medium	High	
Ultraviolet Stability	Low	Medium	High	
Moisture Resistance	Low	Medium	High	
Lubricity	Low	Medium	High	
Electrical Insulation	Low	Medium	High	
Heat Resistance	Low	Medium	High	
Recyclability	Low	Medium	High	
Degradability	Low	Medium	High	
Cost	Low	Medium	High	
Color		Black		
		Gray		
		White		
		Blue		
		Green		
		Red		
		Yellow		

Worksheet 4c: Matching Material Families

List up to three material families that satisfy all the requirements listed in Worksheet 4c.

Part C					
1.		2.		3.	

Worksheet 5c: List of Matching Material Families

Step 6: Determine Matching Materials
Once again, go to the Materials Section and click on the large button called "Properties." This takes you to the main Materials Properties page. Referring to the list you created in Worksheet 3, click the arrow next to the first family. This brings up detailed materials. In the right pane, you see a list of properties such as Hardness, Impact Strength, Rigidity, and so on. Once again select the materials that match the choices in Worksheet 1. List these below in Worksheet 6: Matching Materials. List the matching materials for Part A in Worksheet 6a, the matching materials for Part B in Worksheet 6b, and the matching materials for Part C in Worksheet 6c.

Part A				
Properties	**Value Selected**			**Matching Materials**
Hardness	Low	Medium	High	
Impact Strength	Low	Medium	High	
Rigidity	Low	Medium	High	
Abrasion Resistance	Low	Medium	High	
Optical Clarity	Low	Medium	High	
Chemical Resistance	Low	Medium	High	
Food Compatibility	Low	Medium	High	
Ultraviolet Stability	Low	Medium	High	
Moisture Resistance	Low	Medium	High	
Lubricity	Low	Medium	High	
Electrical Insulation	Low	Medium	High	
Heat Resistance	Low	Medium	High	
Recyclability	Low	Medium	High	
Degradability	Low	Medium	High	
Cost	Low	Medium	High	
Color	Black			
	Gray			
	White			
	Blue			
	Green			
	Red			
	Yellow			

Worksheet 6a: Matching Materials

List up to three materials that best satisfy all the requirements listed in Worksheet 3.

Part A					
1.		2.		3.	

Worksheet 7a: List of Matching Materials

Part B				
Properties	Value Selected			Matching Materials
Hardness	Low	Medium	High	
Impact Strength	Low	Medium	High	
Rigidity	Low	Medium	High	
Abrasion Resistance	Low	Medium	High	
Optical Clarity	Low	Medium	High	
Chemical Resistance	Low	Medium	High	
Food Compatibility	Low	Medium	High	
Ultraviolet Stability	Low	Medium	High	
Moisture Resistance	Low	Medium	High	
Lubricity	Low	Medium	High	
Electrical Insulation	Low	Medium	High	
Heat Resistance	Low	Medium	High	
Recyclability	Low	Medium	High	
Degradability	Low	Medium	High	
Cost	Low	Medium	High	
Color	Black			
	Gray			
	White			
	Blue			
	Green			
	Red			
	Yellow			

Worksheet 6b: Matching Materials

List up to three materials that best satisfy all the requirements listed in Worksheet 3.

Part B					
1.		2.		3.	

Worksheet 7b: List of Matching Materials

Part C				
Properties	**Value Selected**			**Matching Materials**
Hardness	Low	Medium	High	
Impact Strength	Low	Medium	High	
Rigidity	Low	Medium	High	
Abrasion Resistance	Low	Medium	High	
Optical Clarity	Low	Medium	High	
Chemical Resistance	Low	Medium	High	
Food Compatibility	Low	Medium	High	
Ultraviolet Stability	Low	Medium	High	
Moisture Resistance	Low	Medium	High	
Lubricity	Low	Medium	High	
Electrical Insulation	Low	Medium	High	
Heat Resistance	Low	Medium	High	
Recyclability	Low	Medium	High	
Degradability	Low	Medium	High	
Cost	Low	Medium	High	
Color	Black			
	Gray			
	White			
	Blue			
	Green			
	Red			
	Yellow			

Worksheet 6c: Matching Materials

List up to three materials that best satisfy all the requirements listed in Worksheet 3.

Part C		
1.	2.	3.

Worksheet 7c: List of Matching Materials

SELECTING PROCESSES

Step 8: Define the Form Requirements
Study the problem description, define the form characteristics desired of the product, and list these below in Worksheet 8: List of Form Requirements.

Part A
Part B
Part C

Worksheet 8: List of Form Requirements

Step 9: Determine the Stages by which the Desired Form is Achieved
Determine how the form requirements, described in Worksheet 8: List of Form Requirements, can be achieved as a series of successive stages. List each part and the necessary operations required to achieve the desired form in stages. If there are multiple parts in the product, call them out as well. List this information below in Worksheet 9: Stages of Form Generation.

Stage	What form is achieved in this stage?
	Part A
	Part B
	Part C

Worksheet 9: Stages of Form Generation

Step 10: Determine the Manufacturing Processes that are Relevant

Refer to the materials selections listed previously in Worksheet 7: List of Matching Materials, and the form requirements listed in Worksheet 9: Stages of Form Generation. Apply the procedure described below to determine the processes that are appropriate for each stage of processing each of the selected materials.

- Load MaterialTool and on the Title page click on the button called "**Processes.**" This takes you to the Processes section of MaterialTool.
- Click on any of the process groups buttons: "**Finishing**", "**Forming**", "**Joining**" or "**Machining**."
  This takes you to the selected processes group page. Notice that all major material families (such as Ceramic, Leather, and Wood) are listed in the definition. By clicking on these words, it is possible to determine the processes applicable to each material group.
- Click on the button with the name of an individual process. This takes you to that particular process page. By reading the process definition, design tips, and visual consequences, determine if the process can be used to achieve the desired form in the particular material.
- For a deeper understanding of the process, click on the green button named "**Show principle**."
  List the selected processes below in Worksheet 10: List of Selected Manufacturing Processes.

Step	Process	Description
Part A		
1.		
2.		
3.		
4.		
5.		
Part B		
1.		
2.		
3.		
4.		
5.		
Part C		
1.		
2.		
3.		
4.		
5.		

Worksheet 10: List of Selected Manufacturing Processes

FINAL SELECTION OF MATERIALS & PROCESSES

Step 11: List the Final Selections of Materials and Processes
Refer to the materials selections listed previously in Worksheet 7: List of Matching Materials, and the form requirements listed in Worksheet 10: List of Selected Manufacturing Processes. List the final choices of materials and manufacturing processes below in the Final Report.

FINAL REPORT	
Part A	
Material Selected	
Properties	
Rationale	
Processes Selected	
Operation	
Rationale	
Part B	
Material Selected	
Properties	
Rationale	
Processes Selected	
Operation	
Rationale	
Part C	
Material Selected	
Properties	
Rationale	
Processes Selected	
Operation	
Rationale	

Worksheet 11: Final Report

Unit 9: Belt

Problem
The picture shows a typical belt used with trousers.
The belt is manufactured in different lengths but with the same buckle.
The belt needs to be strong, supple, have an attractive polished surface, and be available in a few colors. How many components is the belt made up of, and what materials and processes are used for manufacturing these parts?

SELECTING MATERIALS

Step 1: Identify the Parts of the Product
The product shown above may be made up of more than one part. Examine the photograph of the product, and identify all the parts of the product. List these in Worksheet 1: List of all Product Parts.

1.	
2.	
3.	
4.	
5.	
6.	
7.	
8.	
9.	
10.	

Worksheet 1: List of all Product Parts

Step 2: List Unique Parts
Some of the parts listed above may be identical parts. Identify the unique parts of the product. Consider mirrored parts, such as a left half and a right half of a product, to be identical parts. List up to three unique parts in Worksheet 2: Unique Product Parts.

Part A	
Part B	
Part C	

Worksheet 2: Unique Product Parts

Step 3: Define the Desired Property Values

First identify what performance characteristics are required for each part that has been identified above. Circle the property values for the material that satisfy the performance requirements for each part in Worksheet 3: Property Values below. Note that only some of the properties will be relevant for each part; leave the rest blank.

	Part A						
Hardness	Low		Medium		High		
Impact Strength	Low		Medium		High		
Rigidity	Low		Medium		High		
Abrasion Resistance	Low		Medium		High		
Optical Clarity	Low		Medium		High		
Chemical Resistance	Low		Medium		High		
Food Compatibility	Low		Medium		High		
Ultraviolet Stability	Low		Medium		High		
Moisture Resistance	Low		Medium		High		
Lubricity	Low		Medium		High		
Electrical Insulation	Low		Medium		High		
Heat Resistance	Low		Medium		High		
Recyclability	Low		Medium		High		
Degradability	Low		Medium		High		
Cost	Low		Medium		High		
Color	Black	Gray	White	Blue	Green	Red	Yellow
	Part B						
Hardness	Low		Medium		High		
Impact Strength	Low		Medium		High		
Rigidity	Low		Medium		High		
Abrasion Resistance	Low		Medium		High		
Optical Clarity	Low		Medium		High		
Chemical Resistance	Low		Medium		High		
Food Compatibility	Low		Medium		High		
Ultraviolet Stability	Low		Medium		High		
Moisture Resistance	Low		Medium		High		
Lubricity	Low		Medium		High		
Electrical Insulation	Low		Medium		High		
Heat Resistance	Low		Medium		High		
Recyclability	Low		Medium		High		
Degradability	Low		Medium		High		
Cost	Low		Medium		High		
Color	Black	Gray	White	Blue	Green	Red	Yellow
	Part C						
Hardness	Low		Medium		High		
Impact Strength	Low		Medium		High		
Rigidity	Low		Medium		High		
Abrasion Resistance	Low		Medium		High		
Optical Clarity	Low		Medium		High		
Chemical Resistance	Low		Medium		High		
Food Compatibility	Low		Medium		High		
Ultraviolet Stability	Low		Medium		High		
Moisture Resistance	Low		Medium		High		
Lubricity	Low		Medium		High		
Electrical Insulation	Low		Medium		High		
Heat Resistance	Low		Medium		High		
Recyclability	Low		Medium		High		
Degradability	Low		Medium		High		
Cost	Low		Medium		High		
Color	Black	Gray	White	Blue	Green	Red	Yellow

Worksheet 3: Property Values

Step 4: Determine Matching Material Families

On the MaterialTool Title page click on the button called "Materials." This takes you to the Materials Section. Click on the large button called "Properties." This takes you to the Materials Properties page. Notice the list of material families listed on the left: Ceramic, Cement, Fiber, and so on. In the right pane you see a list of properties such as Hardness, Impact Strength, Rigidity, and so on. Click on the values Low, Medium, or High to determine Material Families that match the choices from Worksheet 3: Property Values. List the matching families for Part A in Worksheet 4a, the matching families for Part B in Worksheet 4b, and the matching families for Part C in Worksheet 4c.

Part A				
Properties	**Value Selected**			**Matching Materials**
Hardness	Low	Medium	High	
Impact Strength	Low	Medium	High	
Rigidity	Low	Medium	High	
Abrasion Resistance	Low	Medium	High	
Optical Clarity	Low	Medium	High	
Chemical Resistance	Low	Medium	High	
Food Compatibility	Low	Medium	High	
Ultraviolet Stability	Low	Medium	High	
Moisture Resistance	Low	Medium	High	
Lubricity	Low	Medium	High	
Electrical Insulation	Low	Medium	High	
Heat Resistance	Low	Medium	High	
Recyclability	Low	Medium	High	
Degradability	Low	Medium	High	
Cost	Low	Medium	High	
Color	Black			
	Gray			
	White			
	Blue			
	Green			
	Red			
	Yellow			

Worksheet 4a: Matching Material Families

Step 5: Identify the Most Appropriate Material Families

List up to three material families that satisfy all the requirements listed in Worksheet 4a.

Part A					
1.		2.		3.	

Worksheet 5a: List of Matching Material Families

Part B				
Properties	**Value Selected**			**Matching Materials**
Hardness	Low	Medium	High	
Impact Strength	Low	Medium	High	
Rigidity	Low	Medium	High	
Abrasion Resistance	Low	Medium	High	
Optical Clarity	Low	Medium	High	
Chemical Resistance	Low	Medium	High	
Food Compatibility	Low	Medium	High	
Ultraviolet Stability	Low	Medium	High	
Moisture Resistance	Low	Medium	High	
Lubricity	Low	Medium	High	
Electrical Insulation	Low	Medium	High	
Heat Resistance	Low	Medium	High	
Recyclability	Low	Medium	High	
Degradability	Low	Medium	High	
Cost	Low	Medium	High	
Color	Black			
	Gray			
	White			
	Blue			
	Green			
	Red			
	Yellow			

Worksheet 4b: Matching Material Families

List up to three material families that satisfy all the requirements listed in Worksheet 4b.

Part B					
1.		2.		3.	

Worksheet 5b: List of Matching Material Families

Part C				
Properties	**Value Selected**			**Matching Materials**
Hardness	Low	Medium	High	
Impact Strength	Low	Medium	High	
Rigidity	Low	Medium	High	
Abrasion Resistance	Low	Medium	High	
Optical Clarity	Low	Medium	High	
Chemical Resistance	Low	Medium	High	
Food Compatibility	Low	Medium	High	
Ultraviolet Stability	Low	Medium	High	
Moisture Resistance	Low	Medium	High	
Lubricity	Low	Medium	High	
Electrical Insulation	Low	Medium	High	
Heat Resistance	Low	Medium	High	
Recyclability	Low	Medium	High	
Degradability	Low	Medium	High	
Cost	Low	Medium	High	
Color	Black			
	Gray			
	White			
	Blue			
	Green			
	Red			
	Yellow			

Worksheet 4c: Matching Material Families

List up to three material families that satisfy all the requirements listed in Worksheet 4c.

Part C		
1.	2.	3.

Worksheet 5c: List of Matching Material Families

Step 6: Determine Matching Materials
Once again, go to the Materials Section and click on the large button called "Properties." This takes you to the main Materials Properties page. Referring to the list you created in Worksheet 3, click the arrow next to the first family. This brings up detailed materials. In the right pane, you see a list of properties such as Hardness, Impact Strength, Rigidity, and so on. Once again select the materials that match the choices in Worksheet 1. List these below in Worksheet 6: Matching Materials. List the matching materials for Part A in Worksheet 6a, the matching materials for Part B in Worksheet 6b, and the matching materials for Part C in Worksheet 6c.

Part A				
Properties	**Value Selected**			**Matching Materials**
Hardness	Low	Medium	High	
Impact Strength	Low	Medium	High	
Rigidity	Low	Medium	High	
Abrasion Resistance	Low	Medium	High	
Optical Clarity	Low	Medium	High	
Chemical Resistance	Low	Medium	High	
Food Compatibility	Low	Medium	High	
Ultraviolet Stability	Low	Medium	High	
Moisture Resistance	Low	Medium	High	
Lubricity	Low	Medium	High	
Electrical Insulation	Low	Medium	High	
Heat Resistance	Low	Medium	High	
Recyclability	Low	Medium	High	
Degradability	Low	Medium	High	
Cost	Low	Medium	High	
Color	Black			
	Gray			
	White			
	Blue			
	Green			
	Red			
	Yellow			

Worksheet 6a: Matching Materials

List up to three materials that best satisfy all the requirements listed in Worksheet 3.

Part A		
1.	2.	3.

Worksheet 7a: List of Matching Materials

Part B				
Properties	**Value Selected**			**Matching Materials**
Hardness	Low	Medium	High	
Impact Strength	Low	Medium	High	
Rigidity	Low	Medium	High	
Abrasion Resistance	Low	Medium	High	
Optical Clarity	Low	Medium	High	
Chemical Resistance	Low	Medium	High	
Food Compatibility	Low	Medium	High	
Ultraviolet Stability	Low	Medium	High	
Moisture Resistance	Low	Medium	High	
Lubricity	Low	Medium	High	
Electrical Insulation	Low	Medium	High	
Heat Resistance	Low	Medium	High	
Recyclability	Low	Medium	High	
Degradability	Low	Medium	High	
Cost	Low	Medium	High	
Color		Black		
		Gray		
		White		
		Blue		
		Green		
		Red		
		Yellow		

Worksheet 6b: Matching Materials

List up to three materials that best satisfy all the requirements listed in Worksheet 3.

Part B					
1.		2.		3.	

Worksheet 7b: List of Matching Materials

Part C				
Properties	**Value Selected**			**Matching Materials**
Hardness	Low	Medium	High	
Impact Strength	Low	Medium	High	
Rigidity	Low	Medium	High	
Abrasion Resistance	Low	Medium	High	
Optical Clarity	Low	Medium	High	
Chemical Resistance	Low	Medium	High	
Food Compatibility	Low	Medium	High	
Ultraviolet Stability	Low	Medium	High	
Moisture Resistance	Low	Medium	High	
Lubricity	Low	Medium	High	
Electrical Insulation	Low	Medium	High	
Heat Resistance	Low	Medium	High	
Recyclability	Low	Medium	High	
Degradability	Low	Medium	High	
Cost	Low	Medium	High	
Color	Black			
	Gray			
	White			
	Blue			
	Green			
	Red			
	Yellow			

Worksheet 6c: Matching Materials

List up to three materials that best satisfy all the requirements listed in Worksheet 3.

Part C		
1.	2.	3.

Worksheet 7c: List of Matching Materials

SELECTING PROCESSES

Step 8: Define the Form Requirements
Study the problem description, define the form characteristics desired of the product, and list these below in Worksheet 8: List of Form Requirements.

Part A

Part B

Part C

Worksheet 8: List of Form Requirements

Step 9: Determine the Stages by which the Desired Form is Achieved
Determine how the form requirements, described in Worksheet 8: List of Form Requirements, can be achieved as a series of successive stages. List each part and the necessary operations required to achieve the desired form in stages. If there are multiple parts in the product, call them out as well. List this information below in Worksheet 9: Stages of Form Generation.

Stage	What form is achieved in this stage?
	Part A
	Part B
	Part C

Worksheet 9: Stages of Form Generation

Step 10: Determine the Manufacturing Processes that are Relevant

Refer to the materials selections listed previously in Worksheet 7: List of Matching Materials, and the form requirements listed in Worksheet 9: Stages of Form Generation. Apply the procedure described below to determine the processes that are appropriate for each stage of processing each of the selected materials.

- Load MaterialTool and on the Title page click on the button called "**Processes.**" This takes you to the Processes section of MaterialTool.
- Click on any of the process groups buttons: "**Finishing**", "**Forming**", "**Joining**" or "**Machining**."
  This takes you to the selected processes group page. Notice that all major material families (such as Ceramic, Leather, and Wood) are listed in the definition. By clicking on these words, it is possible to determine the processes applicable to each material group.
- Click on the button with the name of an individual process. This takes you to that particular process page. By reading the process definition, design tips, and visual consequences, determine if the process can be used to achieve the desired form in the particular material.
- For a deeper understanding of the process, click on the green button named "**Show principle**."
  List the selected processes below in Worksheet 10: List of Selected Manufacturing Processes.

Step	Process	Description
Part A		
1.		
2.		
3.		
4.		
5.		
Part B		
1.		
2.		
3.		
4.		
5.		
Part C		
1.		
2.		
3.		
4.		
5.		

Worksheet 10: List of Selected Manufacturing Processes

FINAL SELECTION OF MATERIALS & PROCESSES

Step 11: List the Final Selections of Materials and Processes

Refer to the materials selections listed previously in Worksheet 7: List of Matching Materials, and the form requirements listed in Worksheet 10: List of Selected Manufacturing Processes. List the final choices of materials and manufacturing processes below in the Final Report.

FINAL REPORT	
Part A	
Material Selected	
Properties	
Rationale	
Processes Selected	
Operation	
Rationale	
Part B	
Material Selected	
Properties	
Rationale	
Processes Selected	
Operation	
Rationale	
Part C	
Material Selected	
Properties	
Rationale	
Processes Selected	
Operation	
Rationale	

Worksheet 11: Final Report

Unit 10: Garden Tool

Problem
What materials and technological processes would you recommend to produce a garden tool similar to the one shown in the picture on the right? In considering your advice keep in mind that this multifunctional garden tool needs to be durable, resistant to corrosion, and inexpensive.

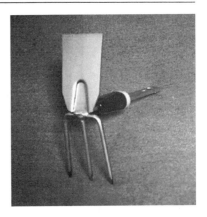

SELECTING MATERIALS

Step 1: Identify the Parts of the Product
The product shown above may be made up of more than one part. Examine the photograph of the product, and identify all the parts of the product. List these in Worksheet 1: List of all Product Parts.

1.	
2.	
3.	
4.	
5.	
6.	
7.	
8.	
9.	
10.	

Worksheet 1: List of all Product Parts

Step 2: List Unique Parts
Some of the parts listed above may be identical parts. Identify the unique parts of the product. Consider mirrored parts, such as a left half and a right half of a product, to be identical parts. List up to three unique parts in Worksheet 2: Unique Product Parts.

Part A	
Part B	
Part C	

Worksheet 2: Unique Product Parts

Step 3: Define the Desired Property Values

First identify what performance characteristics are required for each part that has been identified above. Circle the property values for the material that satisfy the performance requirements for each part in Worksheet 3: Property Values below. Note that only some of the properties will be relevant for each part; leave the rest blank.

Part A

Hardness	Low			Medium		High	
Impact Strength	Low			Medium		High	
Rigidity	Low			Medium		High	
Abrasion Resistance	Low			Medium		High	
Optical Clarity	Low			Medium		High	
Chemical Resistance	Low			Medium		High	
Food Compatibility	Low			Medium		High	
Ultraviolet Stability	Low			Medium		High	
Moisture Resistance	Low			Medium		High	
Lubricity	Low			Medium		High	
Electrical Insulation	Low			Medium		High	
Heat Resistance	Low			Medium		High	
Recyclability	Low			Medium		High	
Degradability	Low			Medium		High	
Cost	Low			Medium		High	
Color	Black	Gray	White	Blue	Green	Red	Yellow

Part B

Hardness	Low			Medium		High	
Impact Strength	Low			Medium		High	
Rigidity	Low			Medium		High	
Abrasion Resistance	Low			Medium		High	
Optical Clarity	Low			Medium		High	
Chemical Resistance	Low			Medium		High	
Food Compatibility	Low			Medium		High	
Ultraviolet Stability	Low			Medium		High	
Moisture Resistance	Low			Medium		High	
Lubricity	Low			Medium		High	
Electrical Insulation	Low			Medium		High	
Heat Resistance	Low			Medium		High	
Recyclability	Low			Medium		High	
Degradability	Low			Medium		High	
Cost	Low			Medium		High	
Color	Black	Gray	White	Blue	Green	Red	Yellow

Part C

Hardness	Low			Medium		High	
Impact Strength	Low			Medium		High	
Rigidity	Low			Medium		High	
Abrasion Resistance	Low			Medium		High	
Optical Clarity	Low			Medium		High	
Chemical Resistance	Low			Medium		High	
Food Compatibility	Low			Medium		High	
Ultraviolet Stability	Low			Medium		High	
Moisture Resistance	Low			Medium		High	
Lubricity	Low			Medium		High	
Electrical Insulation	Low			Medium		High	
Heat Resistance	Low			Medium		High	
Recyclability	Low			Medium		High	
Degradability	Low			Medium		High	
Cost	Low			Medium		High	
Color	Black	Gray	White	Blue	Green	Red	Yellow

Worksheet 3: Property Values

Step 4: Determine Matching Material Families

On the MaterialTool Title page click on the button called "Materials." This takes you to the Materials Section. Click on the large button called "Properties." This takes you to the Materials Properties page. Notice the list of material families listed on the left: Ceramic, Cement, Fiber, and so on. In the right pane you see a list of properties such as Hardness, Impact Strength, Rigidity, and so on. Click on the values Low, Medium, or High to determine Material Families that match the choices from Worksheet 3: Property Values. List the matching families for Part A in Worksheet 4a, the matching families for Part B in Worksheet 4b, and the matching families for Part C in Worksheet 4c.

Part A				
Properties	**Value Selected**			**Matching Materials**
Hardness	Low	Medium	High	
Impact Strength	Low	Medium	High	
Rigidity	Low	Medium	High	
Abrasion Resistance	Low	Medium	High	
Optical Clarity	Low	Medium	High	
Chemical Resistance	Low	Medium	High	
Food Compatibility	Low	Medium	High	
Ultraviolet Stability	Low	Medium	High	
Moisture Resistance	Low	Medium	High	
Lubricity	Low	Medium	High	
Electrical Insulation	Low	Medium	High	
Heat Resistance	Low	Medium	High	
Recyclability	Low	Medium	High	
Degradability	Low	Medium	High	
Cost	Low	Medium	High	
Color	Black			
	Gray			
	White			
	Blue			
	Green			
	Red			
	Yellow			

Worksheet 4a: Matching Material Families

Step 5: Identify the Most Appropriate Material Families

List up to three material families that satisfy all the requirements listed in Worksheet 4a.

Part A					
1.		2.		3.	

Worksheet 5a: List of Matching Material Families

Part B				
Properties	**Value Selected**			**Matching Materials**
Hardness	Low	Medium	High	
Impact Strength	Low	Medium	High	
Rigidity	Low	Medium	High	
Abrasion Resistance	Low	Medium	High	
Optical Clarity	Low	Medium	High	
Chemical Resistance	Low	Medium	High	
Food Compatibility	Low	Medium	High	
Ultraviolet Stability	Low	Medium	High	
Moisture Resistance	Low	Medium	High	
Lubricity	Low	Medium	High	
Electrical Insulation	Low	Medium	High	
Heat Resistance	Low	Medium	High	
Recyclability	Low	Medium	High	
Degradability	Low	Medium	High	
Cost	Low	Medium	High	
Color		Black		
		Gray		
		White		
		Blue		
		Green		
		Red		
		Yellow		

Worksheet 4b: Matching Material Families

List up to three material families that satisfy all the requirements listed in Worksheet 4b.

Part B					
1.		2.		3.	

Worksheet 5b: List of Matching Material Families

Part C				
Properties	**Value Selected**			**Matching Materials**
Hardness	Low	Medium	High	
Impact Strength	Low	Medium	High	
Rigidity	Low	Medium	High	
Abrasion Resistance	Low	Medium	High	
Optical Clarity	Low	Medium	High	
Chemical Resistance	Low	Medium	High	
Food Compatibility	Low	Medium	High	
Ultraviolet Stability	Low	Medium	High	
Moisture Resistance	Low	Medium	High	
Lubricity	Low	Medium	High	
Electrical Insulation	Low	Medium	High	
Heat Resistance	Low	Medium	High	
Recyclability	Low	Medium	High	
Degradability	Low	Medium	High	
Cost	Low	Medium	High	
Color	Black			
	Gray			
	White			
	Blue			
	Green			
	Red			
	Yellow			

Worksheet 4c: Matching Material Families

List up to three material families that satisfy all the requirements listed in Worksheet 4c.

Part C					
1.		2.		3.	

Worksheet 5c: List of Matching Material Families

Step 6: Determine Matching Materials

Once again, go to the Materials Section and click on the large button called "Properties." This takes you to the main Materials Properties page. Referring to the list you created in Worksheet 3, click the arrow next to the first family. This brings up detailed materials. In the right pane, you see a list of properties such as Hardness, Impact Strength, Rigidity, and so on. Once again select the materials that match the choices in Worksheet 1. List these below in Worksheet 6: Matching Materials. List the matching materials for Part A in Worksheet 6a, the matching materials for Part B in Worksheet 6b, and the matching materials for Part C in Worksheet 6c.

Properties	Value Selected			Matching Materials
		Part A		
Hardness	Low	Medium	High	
Impact Strength	Low	Medium	High	
Rigidity	Low	Medium	High	
Abrasion Resistance	Low	Medium	High	
Optical Clarity	Low	Medium	High	
Chemical Resistance	Low	Medium	High	
Food Compatibility	Low	Medium	High	
Ultraviolet Stability	Low	Medium	High	
Moisture Resistance	Low	Medium	High	
Lubricity	Low	Medium	High	
Electrical Insulation	Low	Medium	High	
Heat Resistance	Low	Medium	High	
Recyclability	Low	Medium	High	
Degradability	Low	Medium	High	
Cost	Low	Medium	High	
Color	Black			
	Gray			
	White			
	Blue			
	Green			
	Red			
	Yellow			

Worksheet 6a: Matching Materials

List up to three materials that best satisfy all the requirements listed in Worksheet 3.

Part A		
1.	2.	3.

Worksheet 7a: List of Matching Materials

Part B				
Properties	**Value Selected**			**Matching Materials**
Hardness	Low	Medium	High	
Impact Strength	Low	Medium	High	
Rigidity	Low	Medium	High	
Abrasion Resistance	Low	Medium	High	
Optical Clarity	Low	Medium	High	
Chemical Resistance	Low	Medium	High	
Food Compatibility	Low	Medium	High	
Ultraviolet Stability	Low	Medium	High	
Moisture Resistance	Low	Medium	High	
Lubricity	Low	Medium	High	
Electrical Insulation	Low	Medium	High	
Heat Resistance	Low	Medium	High	
Recyclability	Low	Medium	High	
Degradability	Low	Medium	High	
Cost	Low	Medium	High	
Color	Black			
	Gray			
	White			
	Blue			
	Green			
	Red			
	Yellow			

Worksheet 6b: Matching Materials

List up to three materials that best satisfy all the requirements listed in Worksheet 3.

Part B					
1.		2.		3.	

Worksheet 7b: List of Matching Materials

Part C				
Properties	**Value Selected**			**Matching Materials**
Hardness	Low	Medium	High	
Impact Strength	Low	Medium	High	
Rigidity	Low	Medium	High	
Abrasion Resistance	Low	Medium	High	
Optical Clarity	Low	Medium	High	
Chemical Resistance	Low	Medium	High	
Food Compatibility	Low	Medium	High	
Ultraviolet Stability	Low	Medium	High	
Moisture Resistance	Low	Medium	High	
Lubricity	Low	Medium	High	
Electrical Insulation	Low	Medium	High	
Heat Resistance	Low	Medium	High	
Recyclability	Low	Medium	High	
Degradability	Low	Medium	High	
Cost	Low	Medium	High	
Color	Black			
	Gray			
	White			
	Blue			
	Green			
	Red			
	Yellow			

Worksheet 6c: Matching Materials

List up to three materials that best satisfy all the requirements listed in Worksheet 3.

Part C		
1.	2.	3.

Worksheet 7c: List of Matching Materials

SELECTING PROCESSES

Step 8: Define the Form Requirements
Study the problem description, define the form characteristics desired of the product, and list these below in Worksheet 8: List of Form Requirements.

Part A
Part B
Part C

Worksheet 8: List of Form Requirements

Step 9: Determine the Stages by which the Desired Form is Achieved
Determine how the form requirements, described in Worksheet 8: List of Form Requirements, can be achieved as a series of successive stages. List each part and the necessary operations required to achieve the desired form in stages. If there are multiple parts in the product, call them out as well. List this information below in Worksheet 9: Stages of Form Generation.

Stage	What form is achieved in this stage?
	Part A
	Part B
	Part C

Worksheet 9: Stages of Form Generation

121

Step 10: Determine the Manufacturing Processes that are Relevant
Refer to the materials selections listed previously in Worksheet 7: List of Matching Materials, and the form requirements listed in Worksheet 9: Stages of Form Generation. Apply the procedure described below to determine the processes that are appropriate for each stage of processing each of the selected materials.

- Load MaterialTool and on the Title page click on the button called "**Processes.**" This takes you to the Processes section of MaterialTool.
- Click on any of the process groups buttons: "**Finishing**", "**Forming**", "**Joining**" or "**Machining**." This takes you to the selected processes group page. Notice that all major material families (such as Ceramic, Leather, and Wood) are listed in the definition. By clicking on these words, it is possible to determine the processes applicable to each material group.
- Click on the button with the name of an individual process. This takes you to that particular process page. By reading the process definition, design tips, and visual consequences, determine if the process can be used to achieve the desired form in the particular material.
- For a deeper understanding of the process, click on the green button named "**Show principle**."

List the selected processes below in Worksheet 10: List of Selected Manufacturing Processes.

Step	Process	Description
Part A		
1.		
2.		
3.		
4.		
5.		
Part B		
1.		
2.		
3.		
4.		
5.		
Part C		
1.		
2.		
3.		
4.		
5.		

Worksheet 10: List of Selected Manufacturing Processes

FINAL SELECTION OF MATERIALS & PROCESSES

Step 11: List the Final Selections of Materials and Processes
Refer to the materials selections listed previously in Worksheet 7: List of Matching Materials, and the form requirements listed in Worksheet 10: List of Selected Manufacturing Processes. List the final choices of materials and manufacturing processes below in the Final Report.

FINAL REPORT	
Part A	
Material Selected	
Properties	
Rationale	
Processes Selected	
Operation	
Rationale	
Part B	
Material Selected	
Properties	
Rationale	
Processes Selected	
Operation	
Rationale	
Part C	
Material Selected	
Properties	
Rationale	
Processes Selected	
Operation	
Rationale	

Worksheet 11: Final Report

Unit 11: Salt & Pepper Shakers

Problem
These salt and pepper containers were made from transparent material to ensure that the contents can be seen from the outside. They are sold at a dollar store and are very inexpensive. What specific materials and manufacturing processes were likely used in their production?
What was the sequence of manufacturing processes?

SELECTING MATERIALS

Step 1: Identify the Parts of the Product
The product shown above may be made up of more than one part. Examine the photograph of the product, and identify all the parts of the product. List these in Worksheet 1: List of all Product Parts.

1.	
2.	
3.	
4.	
5.	
6.	
7.	
8.	
9.	
10.	

Worksheet 1: List of all Product Parts

Step 2: List Unique Parts
Some of the parts listed above may be identical parts. Identify the unique parts of the product. Consider mirrored parts, such as a left half and a right half of a product, to be identical parts. List up to three unique parts in Worksheet 2: Unique Product Parts.

Part A	
Part B	
Part C	

Worksheet 2: Unique Product Parts

Step 3: Define the Desired Property Values

First identify what performance characteristics are required for each part that has been identified above. Circle the property values for the material that satisfy the performance requirements for each part in Worksheet 3: Property Values below. Note that only some of the properties will be relevant for each part; leave the rest blank.

Part A							
Hardness	Low		Medium			High	
Impact Strength	Low		Medium			High	
Rigidity	Low		Medium			High	
Abrasion Resistance	Low		Medium			High	
Optical Clarity	Low		Medium			High	
Chemical Resistance	Low		Medium			High	
Food Compatibility	Low		Medium			High	
Ultraviolet Stability	Low		Medium			High	
Moisture Resistance	Low		Medium			High	
Lubricity	Low		Medium			High	
Electrical Insulation	Low		Medium			High	
Heat Resistance	Low		Medium			High	
Recyclability	Low		Medium			High	
Degradability	Low		Medium			High	
Cost	Low		Medium			High	
Color	Black	Gray	White	Blue	Green	Red	Yellow
Part B							
Hardness	Low		Medium			High	
Impact Strength	Low		Medium			High	
Rigidity	Low		Medium			High	
Abrasion Resistance	Low		Medium			High	
Optical Clarity	Low		Medium			High	
Chemical Resistance	Low		Medium			High	
Food Compatibility	Low		Medium			High	
Ultraviolet Stability	Low		Medium			High	
Moisture Resistance	Low		Medium			High	
Lubricity	Low		Medium			High	
Electrical Insulation	Low		Medium			High	
Heat Resistance	Low		Medium			High	
Recyclability	Low		Medium			High	
Degradability	Low		Medium			High	
Cost	Low		Medium			High	
Color	Black	Gray	White	Blue	Green	Red	Yellow
Part C							
Hardness	Low		Medium			High	
Impact Strength	Low		Medium			High	
Rigidity	Low		Medium			High	
Abrasion Resistance	Low		Medium			High	
Optical Clarity	Low		Medium			High	
Chemical Resistance	Low		Medium			High	
Food Compatibility	Low		Medium			High	
Ultraviolet Stability	Low		Medium			High	
Moisture Resistance	Low		Medium			High	
Lubricity	Low		Medium			High	
Electrical Insulation	Low		Medium			High	
Heat Resistance	Low		Medium			High	
Recyclability	Low		Medium			High	
Degradability	Low		Medium			High	
Cost	Low		Medium			High	
Color	Black	Gray	White	Blue	Green	Red	Yellow

Worksheet 3: Property Values

Step 4: Determine Matching Material Families

On the MaterialTool Title page click on the button called "Materials." This takes you to the Materials Section. Click on the large button called "Properties." This takes you to the Materials Properties page. Notice the list of material families listed on the left: Ceramic, Cement, Fiber, and so on. In the right pane you see a list of properties such as Hardness, Impact Strength, Rigidity, and so on. Click on the values Low, Medium, or High to determine Material Families that match the choices from Worksheet 3: Property Values. List the matching families for Part A in Worksheet 4a, the matching families for Part B in Worksheet 4b, and the matching families for Part C in Worksheet 4c.

Part A				
Properties	**Value Selected**			**Matching Materials**
Hardness	Low	Medium	High	
Impact Strength	Low	Medium	High	
Rigidity	Low	Medium	High	
Abrasion Resistance	Low	Medium	High	
Optical Clarity	Low	Medium	High	
Chemical Resistance	Low	Medium	High	
Food Compatibility	Low	Medium	High	
Ultraviolet Stability	Low	Medium	High	
Moisture Resistance	Low	Medium	High	
Lubricity	Low	Medium	High	
Electrical Insulation	Low	Medium	High	
Heat Resistance	Low	Medium	High	
Recyclability	Low	Medium	High	
Degradability	Low	Medium	High	
Cost	Low	Medium	High	
Color	Black			
	Gray			
	White			
	Blue			
	Green			
	Red			
	Yellow			

Worksheet 4a: Matching Material Families

Step 5: Identify the Most Appropriate Material Families

List up to three material families that satisfy all the requirements listed in Worksheet 4a.

Part A		
1.	2.	3.

Worksheet 5a: List of Matching Material Families

Part B				
Properties	**Value Selected**			**Matching Materials**
Hardness	Low	Medium	High	
Impact Strength	Low	Medium	High	
Rigidity	Low	Medium	High	
Abrasion Resistance	Low	Medium	High	
Optical Clarity	Low	Medium	High	
Chemical Resistance	Low	Medium	High	
Food Compatibility	Low	Medium	High	
Ultraviolet Stability	Low	Medium	High	
Moisture Resistance	Low	Medium	High	
Lubricity	Low	Medium	High	
Electrical Insulation	Low	Medium	High	
Heat Resistance	Low	Medium	High	
Recyclability	Low	Medium	High	
Degradability	Low	Medium	High	
Cost	Low	Medium	High	
Color	Black			
	Gray			
	White			
	Blue			
	Green			
	Red			
	Yellow			

Worksheet 4b: Matching Material Families

List up to three material families that satisfy all the requirements listed in Worksheet 4b.

Part B					
1.		2.		3.	

Worksheet 5b: List of Matching Material Families

Part C				
Properties	**Value Selected**			**Matching Materials**
Hardness	Low	Medium	High	
Impact Strength	Low	Medium	High	
Rigidity	Low	Medium	High	
Abrasion Resistance	Low	Medium	High	
Optical Clarity	Low	Medium	High	
Chemical Resistance	Low	Medium	High	
Food Compatibility	Low	Medium	High	
Ultraviolet Stability	Low	Medium	High	
Moisture Resistance	Low	Medium	High	
Lubricity	Low	Medium	High	
Electrical Insulation	Low	Medium	High	
Heat Resistance	Low	Medium	High	
Recyclability	Low	Medium	High	
Degradability	Low	Medium	High	
Cost	Low	Medium	High	
Color	Black			
	Gray			
	White			
	Blue			
	Green			
	Red			
	Yellow			

Worksheet 4c: Matching Material Families

List up to three material families that satisfy all the requirements listed in Worksheet 4c.

Part C					
1.		2.		3.	

Worksheet 5c: List of Matching Material Families

Step 6: Determine Matching Materials
Once again, go to the Materials Section and click on the large button called "Properties." This takes you to the main Materials Properties page. Referring to the list you created in Worksheet 3, click the arrow next to the first family. This brings up detailed materials. In the right pane, you see a list of properties such as Hardness, Impact Strength, Rigidity, and so on. Once again select the materials that match the choices in Worksheet 1. List these below in Worksheet 6: Matching Materials. List the matching materials for Part A in Worksheet 6a, the matching materials for Part B in Worksheet 6b, and the matching materials for Part C in Worksheet 6c.

Part A				
Properties	**Value Selected**			**Matching Materials**
Hardness	Low	Medium	High	
Impact Strength	Low	Medium	High	
Rigidity	Low	Medium	High	
Abrasion Resistance	Low	Medium	High	
Optical Clarity	Low	Medium	High	
Chemical Resistance	Low	Medium	High	
Food Compatibility	Low	Medium	High	
Ultraviolet Stability	Low	Medium	High	
Moisture Resistance	Low	Medium	High	
Lubricity	Low	Medium	High	
Electrical Insulation	Low	Medium	High	
Heat Resistance	Low	Medium	High	
Recyclability	Low	Medium	High	
Degradability	Low	Medium	High	
Cost	Low	Medium	High	
Color		Black		
		Gray		
		White		
		Blue		
		Green		
		Red		
		Yellow		

Worksheet 6a: Matching Materials

List up to three materials that best satisfy all the requirements listed in Worksheet 3.

Part A					
1.		2.		3.	

Worksheet 7a: List of Matching Materials

Part B				
Properties	**Value Selected**			**Matching Materials**
Hardness	Low	Medium	High	
Impact Strength	Low	Medium	High	
Rigidity	Low	Medium	High	
Abrasion Resistance	Low	Medium	High	
Optical Clarity	Low	Medium	High	
Chemical Resistance	Low	Medium	High	
Food Compatibility	Low	Medium	High	
Ultraviolet Stability	Low	Medium	High	
Moisture Resistance	Low	Medium	High	
Lubricity	Low	Medium	High	
Electrical Insulation	Low	Medium	High	
Heat Resistance	Low	Medium	High	
Recyclability	Low	Medium	High	
Degradability	Low	Medium	High	
Cost	Low	Medium	High	
Color	Black			
	Gray			
	White			
	Blue			
	Green			
	Red			
	Yellow			

Worksheet 6b: Matching Materials

List up to three materials that best satisfy all the requirements listed in Worksheet 3.

Part B					
1.		2.		3.	

Worksheet 7b: List of Matching Materials

Part C				
Properties	**Value Selected**			**Matching Materials**
Hardness	Low	Medium	High	
Impact Strength	Low	Medium	High	
Rigidity	Low	Medium	High	
Abrasion Resistance	Low	Medium	High	
Optical Clarity	Low	Medium	High	
Chemical Resistance	Low	Medium	High	
Food Compatibility	Low	Medium	High	
Ultraviolet Stability	Low	Medium	High	
Moisture Resistance	Low	Medium	High	
Lubricity	Low	Medium	High	
Electrical Insulation	Low	Medium	High	
Heat Resistance	Low	Medium	High	
Recyclability	Low	Medium	High	
Degradability	Low	Medium	High	
Cost	Low	Medium	High	
Color		Black		
		Gray		
		White		
		Blue		
		Green		
		Red		
		Yellow		

Worksheet 6c: Matching Materials

List up to three materials that best satisfy all the requirements listed in Worksheet 3.

Part C				
1.		2.	3.	

Worksheet 7c: List of Matching Materials

SELECTING PROCESSES

Step 8: Define the Form Requirements
Study the problem description, define the form characteristics desired of the product, and list these below in Worksheet 8: List of Form Requirements.

Part A
Part B
Part C

Worksheet 8: List of Form Requirements

Step 9: Determine the Stages by which the Desired Form is Achieved
Determine how the form requirements, described in Worksheet 8: List of Form Requirements, can be achieved as a series of successive stages. List each part and the necessary operations required to achieve the desired form in stages. If there are multiple parts in the product, call them out as well. List this information below in Worksheet 9: Stages of Form Generation.

Stage	What form is achieved in this stage?
Part A	
Part B	
Part C	

Worksheet 9: Stages of Form Generation

Step 10: Determine the Manufacturing Processes that are Relevant

Refer to the materials selections listed previously in Worksheet 7: List of Matching Materials, and the form requirements listed in Worksheet 9: Stages of Form Generation. Apply the procedure described below to determine the processes that are appropriate for each stage of processing each of the selected materials.

- Load MaterialTool and on the Title page click on the button called "**Processes.**" This takes you to the Processes section of MaterialTool.
- Click on any of the process groups buttons: "**Finishing**", "**Forming**", "**Joining**" or "**Machining**."
  This takes you to the selected processes group page. Notice that all major material families (such as Ceramic, Leather, and Wood) are listed in the definition. By clicking on these words, it is possible to determine the processes applicable to each material group.
- Click on the button with the name of an individual process. This takes you to that particular process page. By reading the process definition, design tips, and visual consequences, determine if the process can be used to achieve the desired form in the particular material.
- For a deeper understanding of the process, click on the green button named "**Show principle**."
  List the selected processes below in Worksheet 10: List of Selected Manufacturing Processes.

Step	Process	Description
Part A		
1.		
2.		
3.		
4.		
5.		
Part B		
1.		
2.		
3.		
4.		
5.		
Part C		
1.		
2.		
3.		
4.		
5.		

Worksheet 10: List of Selected Manufacturing Processes

FINAL SELECTION OF MATERIALS & PROCESSES

Step 11: List the Final Selections of Materials and Processes
Refer to the materials selections listed previously in Worksheet 7: List of Matching Materials, and the form requirements listed in Worksheet 10: List of Selected Manufacturing Processes. List the final choices of materials and manufacturing processes below in the Final Report.

FINAL REPORT	
Part A	
Material Selected	
Properties	
Rationale	
Processes Selected	
Operation	
Rationale	
Part B	
Material Selected	
Properties	
Rationale	
Processes Selected	
Operation	
Rationale	
Part C	
Material Selected	
Properties	
Rationale	
Processes Selected	
Operation	
Rationale	

Worksheet 11: Final Report

Unit 12: Bottle Opener

Problem
This bottle opener is a multipurpose product designed to remove bottle caps and pierce juice cans. It is a very inexpensive, mass-produced product. What materials is it probably made of? What technologies were employed in its production? What was the sequence of manufacturing processes?

SELECTING MATERIALS

Step 1: Identify the Parts of the Product
The product shown above may be made up of more than one part. Examine the photograph of the product, and identify all the parts of the product. List these in Worksheet 1: List of all Product Parts.

1.	
2.	
3.	
4.	
5.	
6.	
7.	
8.	
9.	
10.	

Worksheet 1: List of all Product Parts

Step 2: List Unique Parts
Some of the parts listed above may be identical parts. Identify the unique parts of the product. Consider mirrored parts, such as a left half and a right half of a product, to be identical parts. List up to three unique parts in Worksheet 2: Unique Product Parts.

Part A	
Part B	
Part C	

Worksheet 2: Unique Product Parts

Step 3: Define the Desired Property Values

First identify what performance characteristics are required for each part that has been identified above. Circle the property values for the material that satisfy the performance requirements for each part in Worksheet 3: Property Values below. Note that only some of the properties will be relevant for each part; leave the rest blank.

Part A

Hardness	Low		Medium			High	
Impact Strength	Low		Medium			High	
Rigidity	Low		Medium			High	
Abrasion Resistance	Low		Medium			High	
Optical Clarity	Low		Medium			High	
Chemical Resistance	Low		Medium			High	
Food Compatibility	Low		Medium			High	
Ultraviolet Stability	Low		Medium			High	
Moisture Resistance	Low		Medium			High	
Lubricity	Low		Medium			High	
Electrical Insulation	Low		Medium			High	
Heat Resistance	Low		Medium			High	
Recyclability	Low		Medium			High	
Degradability	Low		Medium			High	
Cost	Low		Medium			High	
Color	Black	Gray	White	Blue	Green	Red	Yellow

Part B

Hardness	Low		Medium			High	
Impact Strength	Low		Medium			High	
Rigidity	Low		Medium			High	
Abrasion Resistance	Low		Medium			High	
Optical Clarity	Low		Medium			High	
Chemical Resistance	Low		Medium			High	
Food Compatibility	Low		Medium			High	
Ultraviolet Stability	Low		Medium			High	
Moisture Resistance	Low		Medium			High	
Lubricity	Low		Medium			High	
Electrical Insulation	Low		Medium			High	
Heat Resistance	Low		Medium			High	
Recyclability	Low		Medium			High	
Degradability	Low		Medium			High	
Cost	Low		Medium			High	
Color	Black	Gray	White	Blue	Green	Red	Yellow

Part C

Hardness	Low		Medium			High	
Impact Strength	Low		Medium			High	
Rigidity	Low		Medium			High	
Abrasion Resistance	Low		Medium			High	
Optical Clarity	Low		Medium			High	
Chemical Resistance	Low		Medium			High	
Food Compatibility	Low		Medium			High	
Ultraviolet Stability	Low		Medium			High	
Moisture Resistance	Low		Medium			High	
Lubricity	Low		Medium			High	
Electrical Insulation	Low		Medium			High	
Heat Resistance	Low		Medium			High	
Recyclability	Low		Medium			High	
Degradability	Low		Medium			High	
Cost	Low		Medium			High	
Color	Black	Gray	White	Blue	Green	Red	Yellow

Worksheet 3: Property Values

Step 4: Determine Matching Material Families
On the MaterialTool Title page click on the button called "Materials." This takes you to the Materials Section. Click on the large button called "Properties." This takes you to the Materials Properties page. Notice the list of material families listed on the left: Ceramic, Cement, Fiber, and so on. In the right pane you see a list of properties such as Hardness, Impact Strength, Rigidity, and so on. Click on the values Low, Medium, or High to determine Material Families that match the choices from Worksheet 3: Property Values. List the matching families for Part A in Worksheet 4a, the matching families for Part B in Worksheet 4b, and the matching families for Part C in Worksheet 4c.

Part A				
Properties	**Value Selected**			**Matching Materials**
Hardness	Low	Medium	High	
Impact Strength	Low	Medium	High	
Rigidity	Low	Medium	High	
Abrasion Resistance	Low	Medium	High	
Optical Clarity	Low	Medium	High	
Chemical Resistance	Low	Medium	High	
Food Compatibility	Low	Medium	High	
Ultraviolet Stability	Low	Medium	High	
Moisture Resistance	Low	Medium	High	
Lubricity	Low	Medium	High	
Electrical Insulation	Low	Medium	High	
Heat Resistance	Low	Medium	High	
Recyclability	Low	Medium	High	
Degradability	Low	Medium	High	
Cost	Low	Medium	High	
Color	Black			
	Gray			
	White			
	Blue			
	Green			
	Red			
	Yellow			

Worksheet 4a: Matching Material Families

Step 5: Identify the Most Appropriate Material Families
List up to three material families that satisfy all the requirements listed in Worksheet 4a.

Part A					
1.		2.		3.	

Worksheet 5a: List of Matching Material Families

Part B				
Properties	**Value Selected**			**Matching Materials**
Hardness	Low	Medium	High	
Impact Strength	Low	Medium	High	
Rigidity	Low	Medium	High	
Abrasion Resistance	Low	Medium	High	
Optical Clarity	Low	Medium	High	
Chemical Resistance	Low	Medium	High	
Food Compatibility	Low	Medium	High	
Ultraviolet Stability	Low	Medium	High	
Moisture Resistance	Low	Medium	High	
Lubricity	Low	Medium	High	
Electrical Insulation	Low	Medium	High	
Heat Resistance	Low	Medium	High	
Recyclability	Low	Medium	High	
Degradability	Low	Medium	High	
Cost	Low	Medium	High	
Color	Black			
	Gray			
	White			
	Blue			
	Green			
	Red			
	Yellow			

Worksheet 4b: Matching Material Families

List up to three material families that satisfy all the requirements listed in Worksheet 4b.

Part B					
1.		2.		3.	

Worksheet 5b: List of Matching Material Families

Part C				
Properties	**Value Selected**			**Matching Materials**
Hardness	Low	Medium	High	
Impact Strength	Low	Medium	High	
Rigidity	Low	Medium	High	
Abrasion Resistance	Low	Medium	High	
Optical Clarity	Low	Medium	High	
Chemical Resistance	Low	Medium	High	
Food Compatibility	Low	Medium	High	
Ultraviolet Stability	Low	Medium	High	
Moisture Resistance	Low	Medium	High	
Lubricity	Low	Medium	High	
Electrical Insulation	Low	Medium	High	
Heat Resistance	Low	Medium	High	
Recyclability	Low	Medium	High	
Degradability	Low	Medium	High	
Cost	Low	Medium	High	
Color		Black		
		Gray		
		White		
		Blue		
		Green		
		Red		
		Yellow		

Worksheet 4c: Matching Material Families

List up to three material families that satisfy all the requirements listed in Worksheet 4c.

Part C		
1.	2.	3.

Worksheet 5c: List of Matching Material Families

Step 6: Determine Matching Materials
Once again, go to the Materials Section and click on the large button called "Properties." This takes you to the main Materials Properties page. Referring to the list you created in Worksheet 3, click the arrow next to the first family. This brings up detailed materials. In the right pane, you see a list of properties such as Hardness, Impact Strength, Rigidity, and so on. Once again select the materials that match the choices in Worksheet 1. List these below in Worksheet 6: Matching Materials. List the matching materials for Part A in Worksheet 6a, the matching materials for Part B in Worksheet 6b, and the matching materials for Part C in Worksheet 6c.

Part A				
Properties	**Value Selected**			**Matching Materials**
Hardness	Low	Medium	High	
Impact Strength	Low	Medium	High	
Rigidity	Low	Medium	High	
Abrasion Resistance	Low	Medium	High	
Optical Clarity	Low	Medium	High	
Chemical Resistance	Low	Medium	High	
Food Compatibility	Low	Medium	High	
Ultraviolet Stability	Low	Medium	High	
Moisture Resistance	Low	Medium	High	
Lubricity	Low	Medium	High	
Electrical Insulation	Low	Medium	High	
Heat Resistance	Low	Medium	High	
Recyclability	Low	Medium	High	
Degradability	Low	Medium	High	
Cost	Low	Medium	High	
Color	Black			
	Gray			
	White			
	Blue			
	Green			
	Red			
	Yellow			

Worksheet 6a: Matching Materials

List up to three materials that best satisfy all the requirements listed in Worksheet 3.

Part A		
1.	2.	3.

Worksheet 7a: List of Matching Materials

Part B				
Properties	**Value Selected**			**Matching Materials**
Hardness	Low	Medium	High	
Impact Strength	Low	Medium	High	
Rigidity	Low	Medium	High	
Abrasion Resistance	Low	Medium	High	
Optical Clarity	Low	Medium	High	
Chemical Resistance	Low	Medium	High	
Food Compatibility	Low	Medium	High	
Ultraviolet Stability	Low	Medium	High	
Moisture Resistance	Low	Medium	High	
Lubricity	Low	Medium	High	
Electrical Insulation	Low	Medium	High	
Heat Resistance	Low	Medium	High	
Recyclability	Low	Medium	High	
Degradability	Low	Medium	High	
Cost	Low	Medium	High	
Color		Black		
		Gray		
		White		
		Blue		
		Green		
		Red		
		Yellow		

Worksheet 6b: Matching Materials

List up to three materials that best satisfy all the requirements listed in Worksheet 3.

Part B					
1.		2.		3.	

Worksheet 7b: List of Matching Materials

Part C				
Properties	**Value Selected**			**Matching Materials**
Hardness	Low	Medium	High	
Impact Strength	Low	Medium	High	
Rigidity	Low	Medium	High	
Abrasion Resistance	Low	Medium	High	
Optical Clarity	Low	Medium	High	
Chemical Resistance	Low	Medium	High	
Food Compatibility	Low	Medium	High	
Ultraviolet Stability	Low	Medium	High	
Moisture Resistance	Low	Medium	High	
Lubricity	Low	Medium	High	
Electrical Insulation	Low	Medium	High	
Heat Resistance	Low	Medium	High	
Recyclability	Low	Medium	High	
Degradability	Low	Medium	High	
Cost	Low	Medium	High	
Color	Black			
	Gray			
	White			
	Blue			
	Green			
	Red			
	Yellow			

Worksheet 6c: Matching Materials

List up to three materials that best satisfy all the requirements listed in Worksheet 3.

Part C		
1.	2.	3.

Worksheet 7c: List of Matching Materials

SELECTING PROCESSES

Step 8: Define the Form Requirements
Study the problem description, define the form characteristics desired of the product, and list these below in Worksheet 8: List of Form Requirements.

Part A

Part B

Part C

Worksheet 8: List of Form Requirements

Step 9: Determine the Stages by which the Desired Form is Achieved
Determine how the form requirements, described in Worksheet 8: List of Form Requirements, can be achieved as a series of successive stages. List each part and the necessary operations required to achieve the desired form in stages. If there are multiple parts in the product, call them out as well. List this information below in Worksheet 9: Stages of Form Generation.

Stage	What form is achieved in this stage?
	Part A
	Part B
	Part C

Worksheet 9: Stages of Form Generation

Step 10: Determine the Manufacturing Processes that are Relevant
Refer to the materials selections listed previously in Worksheet 7: List of Matching Materials, and the form requirements listed in Worksheet 9: Stages of Form Generation. Apply the procedure described below to determine the processes that are appropriate for each stage of processing each of the selected materials.

- Load MaterialTool and on the Title page click on the button called "**Processes.**" This takes you to the Processes section of MaterialTool.
- Click on any of the process groups buttons: "**Finishing**", "**Forming**", "**Joining**" or "**Machining.**"
  This takes you to the selected processes group page. Notice that all major material families (such as Ceramic, Leather, and Wood) are listed in the definition. By clicking on these words, it is possible to determine the processes applicable to each material group.
- Click on the button with the name of an individual process. This takes you to that particular process page. By reading the process definition, design tips, and visual consequences, determine if the process can be used to achieve the desired form in the particular material.
- For a deeper understanding of the process, click on the green button named "**Show principle.**"
  List the selected processes below in Worksheet 10: List of Selected Manufacturing Processes.

Step	Process	Description
Part A		
1.		
2.		
3.		
4.		
5.		
Part B		
1.		
2.		
3.		
4.		
5.		
Part C		
1.		
2.		
3.		
4.		
5.		

Worksheet 10: List of Selected Manufacturing Processes

FINAL SELECTION OF MATERIALS & PROCESSES

Step 11: List the Final Selections of Materials and Processes

Refer to the materials selections listed previously in **Worksheet 7: List of Matching Materials**, and the form requirements listed in **Worksheet 10: List of Selected Manufacturing Processes**. List the final choices of materials and manufacturing processes below in the Final Report.

FINAL REPORT	
Part A	
Material Selected	
Properties	
Rationale	
Processes Selected	
Operation	
Rationale	
Part B	
Material Selected	
Properties	
Rationale	
Processes Selected	
Operation	
Rationale	
Part C	
Material Selected	
Properties	
Rationale	
Processes Selected	
Operation	
Rationale	

Worksheet 11: Final Report

Unit 13: Garlic Press

Problem
This garlic press was designed specifically with two objectives in mind: durability and ease of use. The designer selected two materials for this product. What were her specific material choices? What technologies were used in the production of this garlic press? In what order were they employed?

SELECTING MATERIALS

Step 1: Identify the Parts of the Product
The product shown above may be made up of more than one part. Examine the photograph of the product, and identify all the parts of the product. List these in Worksheet 1: List of all Product Parts.

1.	
2.	
3.	
4.	
5.	
6.	
7.	
8.	
9.	
10.	

Worksheet 1: List of all Product Parts

Step 2: List Unique Parts
Some of the parts listed above may be identical parts. Identify the unique parts of the product. Consider mirrored parts, such as a left half and a right half of a product, to be identical parts. List up to three unique parts in Worksheet 2: Unique Product Parts.

Part A	
Part B	
Part C	

Worksheet 2: Unique Product Parts

146

Step 3: Define the Desired Property Values

First identify what performance characteristics are required for each part that has been identified above. Circle the property values for the material that satisfy the performance requirements for each part in Worksheet 3: Property Values below. Note that only some of the properties will be relevant for each part; leave the rest blank.

Part A							
Hardness	Low		Medium			High	
Impact Strength	Low		Medium			High	
Rigidity	Low		Medium			High	
Abrasion Resistance	Low		Medium			High	
Optical Clarity	Low		Medium			High	
Chemical Resistance	Low		Medium			High	
Food Compatibility	Low		Medium			High	
Ultraviolet Stability	Low		Medium			High	
Moisture Resistance	Low		Medium			High	
Lubricity	Low		Medium			High	
Electrical Insulation	Low		Medium			High	
Heat Resistance	Low		Medium			High	
Recyclability	Low		Medium			High	
Degradability	Low		Medium			High	
Cost	Low		Medium			High	
Color	Black	Gray	White	Blue	Green	Red	Yellow
Part B							
Hardness	Low		Medium			High	
Impact Strength	Low		Medium			High	
Rigidity	Low		Medium			High	
Abrasion Resistance	Low		Medium			High	
Optical Clarity	Low		Medium			High	
Chemical Resistance	Low		Medium			High	
Food Compatibility	Low		Medium			High	
Ultraviolet Stability	Low		Medium			High	
Moisture Resistance	Low		Medium			High	
Lubricity	Low		Medium			High	
Electrical Insulation	Low		Medium			High	
Heat Resistance	Low		Medium			High	
Recyclability	Low		Medium			High	
Degradability	Low		Medium			High	
Cost	Low		Medium			High	
Color	Black	Gray	White	Blue	Green	Red	Yellow
Part C							
Hardness	Low		Medium			High	
Impact Strength	Low		Medium			High	
Rigidity	Low		Medium			High	
Abrasion Resistance	Low		Medium			High	
Optical Clarity	Low		Medium			High	
Chemical Resistance	Low		Medium			High	
Food Compatibility	Low		Medium			High	
Ultraviolet Stability	Low		Medium			High	
Moisture Resistance	Low		Medium			High	
Lubricity	Low		Medium			High	
Electrical Insulation	Low		Medium			High	
Heat Resistance	Low		Medium			High	
Recyclability	Low		Medium			High	
Degradability	Low		Medium			High	
Cost	Low		Medium			High	
Color	Black	Gray	White	Blue	Green	Red	Yellow

Worksheet 3: Property Values

Step 4: Determine Matching Material Families

On the MaterialTool Title page click on the button called "Materials." This takes you to the Materials Section. Click on the large button called "Properties." This takes you to the Materials Properties page. Notice the list of material families listed on the left: Ceramic, Cement, Fiber, and so on. In the right pane you see a list of properties such as Hardness, Impact Strength, Rigidity, and so on. Click on the values Low, Medium, or High to determine Material Families that match the choices from Worksheet 3: Property Values. List the matching families for Part A in Worksheet 4a, the matching families for Part B in Worksheet 4b, and the matching families for Part C in Worksheet 4c.

Part A				
Properties	**Value Selected**			**Matching Materials**
Hardness	Low	Medium	High	
Impact Strength	Low	Medium	High	
Rigidity	Low	Medium	High	
Abrasion Resistance	Low	Medium	High	
Optical Clarity	Low	Medium	High	
Chemical Resistance	Low	Medium	High	
Food Compatibility	Low	Medium	High	
Ultraviolet Stability	Low	Medium	High	
Moisture Resistance	Low	Medium	High	
Lubricity	Low	Medium	High	
Electrical Insulation	Low	Medium	High	
Heat Resistance	Low	Medium	High	
Recyclability	Low	Medium	High	
Degradability	Low	Medium	High	
Cost	Low	Medium	High	
Color	Black			
	Gray			
	White			
	Blue			
	Green			
	Red			
	Yellow			

Worksheet 4a: Matching Material Families

Step 5: Identify the Most Appropriate Material Families

List up to three material families that satisfy all the requirements listed in Worksheet 4a.

Part A		
1.	2.	3.

Worksheet 5a: List of Matching Material Families

Part B				
Properties	**Value Selected**			**Matching Materials**
Hardness	Low	Medium	High	
Impact Strength	Low	Medium	High	
Rigidity	Low	Medium	High	
Abrasion Resistance	Low	Medium	High	
Optical Clarity	Low	Medium	High	
Chemical Resistance	Low	Medium	High	
Food Compatibility	Low	Medium	High	
Ultraviolet Stability	Low	Medium	High	
Moisture Resistance	Low	Medium	High	
Lubricity	Low	Medium	High	
Electrical Insulation	Low	Medium	High	
Heat Resistance	Low	Medium	High	
Recyclability	Low	Medium	High	
Degradability	Low	Medium	High	
Cost	Low	Medium	High	
Color	Black			
	Gray			
	White			
	Blue			
	Green			
	Red			
	Yellow			

Worksheet 4b: Matching Material Families

List up to three material families that satisfy all the requirements listed in Worksheet 4b.

Part B		
1.	2.	3.

Worksheet 5b: List of Matching Material Families

Part C				
Properties	**Value Selected**			**Matching Materials**
Hardness	Low	Medium	High	
Impact Strength	Low	Medium	High	
Rigidity	Low	Medium	High	
Abrasion Resistance	Low	Medium	High	
Optical Clarity	Low	Medium	High	
Chemical Resistance	Low	Medium	High	
Food Compatibility	Low	Medium	High	
Ultraviolet Stability	Low	Medium	High	
Moisture Resistance	Low	Medium	High	
Lubricity	Low	Medium	High	
Electrical Insulation	Low	Medium	High	
Heat Resistance	Low	Medium	High	
Recyclability	Low	Medium	High	
Degradability	Low	Medium	High	
Cost	Low	Medium	High	
Color	Black			
	Gray			
	White			
	Blue			
	Green			
	Red			
	Yellow			

Worksheet 4c: Matching Material Families

List up to three material families that satisfy all the requirements listed in Worksheet 4c.

Part C				
1.		2.		3.

Worksheet 5c: List of Matching Material Families

Step 6: Determine Matching Materials

Once again, go to the Materials Section and click on the large button called "Properties." This takes you to the main Materials Properties page. Referring to the list you created in Worksheet 3, click the arrow next to the first family. This brings up detailed materials. In the right pane, you see a list of properties such as Hardness, Impact Strength, Rigidity, and so on. Once again select the materials that match the choices in Worksheet 1. List these below in Worksheet 6: Matching Materials. List the matching materials for Part A in Worksheet 6a, the matching materials for Part B in Worksheet 6b, and the matching materials for Part C in Worksheet 6c.

Part A				
Properties	**Value Selected**			**Matching Materials**
Hardness	Low	Medium	High	
Impact Strength	Low	Medium	High	
Rigidity	Low	Medium	High	
Abrasion Resistance	Low	Medium	High	
Optical Clarity	Low	Medium	High	
Chemical Resistance	Low	Medium	High	
Food Compatibility	Low	Medium	High	
Ultraviolet Stability	Low	Medium	High	
Moisture Resistance	Low	Medium	High	
Lubricity	Low	Medium	High	
Electrical Insulation	Low	Medium	High	
Heat Resistance	Low	Medium	High	
Recyclability	Low	Medium	High	
Degradability	Low	Medium	High	
Cost	Low	Medium	High	
Color	Black			
	Gray			
	White			
	Blue			
	Green			
	Red			
	Yellow			

Worksheet 6a: Matching Materials

List up to three materials that best satisfy all the requirements listed in Worksheet 3.

Part A		
1.	2.	3.

Worksheet 7a: List of Matching Materials

151

Part B				
Properties	**Value Selected**			**Matching Materials**
Hardness	Low	Medium	High	
Impact Strength	Low	Medium	High	
Rigidity	Low	Medium	High	
Abrasion Resistance	Low	Medium	High	
Optical Clarity	Low	Medium	High	
Chemical Resistance	Low	Medium	High	
Food Compatibility	Low	Medium	High	
Ultraviolet Stability	Low	Medium	High	
Moisture Resistance	Low	Medium	High	
Lubricity	Low	Medium	High	
Electrical Insulation	Low	Medium	High	
Heat Resistance	Low	Medium	High	
Recyclability	Low	Medium	High	
Degradability	Low	Medium	High	
Cost	Low	Medium	High	
Color	Black			
	Gray			
	White			
	Blue			
	Green			
	Red			
	Yellow			

Worksheet 6b: Matching Materials

List up to three materials that best satisfy all the requirements listed in Worksheet 3.

Part B		
1.	2.	3.

Worksheet 7b: List of Matching Materials

Part C				
Properties	**Value Selected**			**Matching Materials**
Hardness	Low	Medium	High	
Impact Strength	Low	Medium	High	
Rigidity	Low	Medium	High	
Abrasion Resistance	Low	Medium	High	
Optical Clarity	Low	Medium	High	
Chemical Resistance	Low	Medium	High	
Food Compatibility	Low	Medium	High	
Ultraviolet Stability	Low	Medium	High	
Moisture Resistance	Low	Medium	High	
Lubricity	Low	Medium	High	
Electrical Insulation	Low	Medium	High	
Heat Resistance	Low	Medium	High	
Recyclability	Low	Medium	High	
Degradability	Low	Medium	High	
Cost	Low	Medium	High	
Color	Black			
	Gray			
	White			
	Blue			
	Green			
	Red			
	Yellow			

Worksheet 6c: Matching Materials

List up to three materials that best satisfy all the requirements listed in Worksheet 3.

Part C			
1.		3.	

Worksheet 7c: List of Matching Materials

SELECTING PROCESSES

Step 8: Define the Form Requirements
Study the problem description, define the form characteristics desired of the product, and list these below in Worksheet 8: List of Form Requirements.

Part A

Part B

Part C

Worksheet 8: List of Form Requirements

Step 9: Determine the Stages by which the Desired Form is Achieved
Determine how the form requirements, described in Worksheet 8: List of Form Requirements, can be achieved as a series of successive stages. List each part and the necessary operations required to achieve the desired form in stages. If there are multiple parts in the product, call them out as well. List this information below in Worksheet 9: Stages of Form Generation.

Stage	What form is achieved in this stage?
Part A	
Part B	
Part C	

Worksheet 9: Stages of Form Generation

Step 10: Determine the Manufacturing Processes that are Relevant
Refer to the materials selections listed previously in Worksheet 7: List of Matching Materials, and the form requirements listed in Worksheet 9: Stages of Form Generation. Apply the procedure described below to determine the processes that are appropriate for each stage of processing each of the selected materials.
- Load MaterialTool and on the Title page click on the button called "**Processes.**" This takes you to the Processes section of MaterialTool.
- Click on any of the process groups buttons: "**Finishing**", "**Forming**", "**Joining**" or "**Machining**."
  This takes you to the selected processes group page. Notice that all major material families (such as Ceramic, Leather, and Wood) are listed in the definition. By clicking on these words, it is possible to determine the processes applicable to each material group.
- Click on the button with the name of an individual process. This takes you to that particular process page. By reading the process definition, design tips, and visual consequences, determine if the process can be used to achieve the desired form in the particular material.
- For a deeper understanding of the process, click on the green button named "**Show principle**."
  List the selected processes below in Worksheet 10: List of Selected Manufacturing Processes.

Step	Process	Description
Part A		
1.		
2.		
3.		
4.		
5.		
Part B		
1.		
2.		
3.		
4.		
5.		
Part C		
1.		
2.		
3.		
4.		
5.		

Worksheet 10: List of Selected Manufacturing Processes

FINAL SELECTION OF MATERIALS & PROCESSES

Step 11: List the Final Selections of Materials and Processes
Refer to the materials selections listed previously in Worksheet 7: List of Matching Materials, and the form requirements listed in Worksheet 10: List of Selected Manufacturing Processes. List the final choices of materials and manufacturing processes below in the Final Report.

FINAL REPORT	
Part A	
Material Selected	
Properties	
Rationale	
Processes Selected	
Operation	
Rationale	
Part B	
Material Selected	
Properties	
Rationale	
Processes Selected	
Operation	
Rationale	
Part C	
Material Selected	
Properties	
Rationale	
Processes Selected	
Operation	
Rationale	

Worksheet 11: Final Report

Unit 14: Baby Bottle

Problem
What materials and technological processes would you recommend to produce a baby bottle set similar to the one shown in the picture on the right? You are asked to offer advice in regard to the most suitable way to apply graphics on the bottle. In considering your advice keep in mind that this bottle needs to have hygienic and aesthetic value, and to be light and durable.

SELECTING MATERIALS

Step 1: Identify the Parts of the Product
The product shown above may be made up of more than one part. Examine the photograph of the product, and identify all the parts of the product. List these in Worksheet 1: List of all Product Parts.

1.	
2.	
3.	
4.	
5.	
6.	
7.	
8.	
9.	
10.	

Worksheet 1: List of all Product Parts

Step 2: List Unique Parts
Some of the parts listed above may be identical parts. Identify the unique parts of the product. Consider mirrored parts, such as a left half and a right half of a product, to be identical parts. List up to three unique parts in Worksheet 2: Unique Product Parts.

Part A	
Part B	
Part C	

Worksheet 2: Unique Product Parts

Step 3: Define the Desired Property Values

First identify what performance characteristics are required for each part that has been identified above. Circle the property values for the material that satisfy the performance requirements for each part in Worksheet 3: Property Values below. Note that only some of the properties will be relevant for each part; leave the rest blank.

Part A							
Hardness	Low		Medium			High	
Impact Strength	Low		Medium			High	
Rigidity	Low		Medium			High	
Abrasion Resistance	Low		Medium			High	
Optical Clarity	Low		Medium			High	
Chemical Resistance	Low		Medium			High	
Food Compatibility	Low		Medium			High	
Ultraviolet Stability	Low		Medium			High	
Moisture Resistance	Low		Medium			High	
Lubricity	Low		Medium			High	
Electrical Insulation	Low		Medium			High	
Heat Resistance	Low		Medium			High	
Recyclability	Low		Medium			High	
Degradability	Low		Medium			High	
Cost	Low		Medium			High	
Color	Black	Gray	White	Blue	Green	Red	Yellow
Part B							
Hardness	Low		Medium			High	
Impact Strength	Low		Medium			High	
Rigidity	Low		Medium			High	
Abrasion Resistance	Low		Medium			High	
Optical Clarity	Low		Medium			High	
Chemical Resistance	Low		Medium			High	
Food Compatibility	Low		Medium			High	
Ultraviolet Stability	Low		Medium			High	
Moisture Resistance	Low		Medium			High	
Lubricity	Low		Medium			High	
Electrical Insulation	Low		Medium			High	
Heat Resistance	Low		Medium			High	
Recyclability	Low		Medium			High	
Degradability	Low		Medium			High	
Cost	Low		Medium			High	
Color	Black	Gray	White	Blue	Green	Red	Yellow
Part C							
Hardness	Low		Medium			High	
Impact Strength	Low		Medium			High	
Rigidity	Low		Medium			High	
Abrasion Resistance	Low		Medium			High	
Optical Clarity	Low		Medium			High	
Chemical Resistance	Low		Medium			High	
Food Compatibility	Low		Medium			High	
Ultraviolet Stability	Low		Medium			High	
Moisture Resistance	Low		Medium			High	
Lubricity	Low		Medium			High	
Electrical Insulation	Low		Medium			High	
Heat Resistance	Low		Medium			High	
Recyclability	Low		Medium			High	
Degradability	Low		Medium			High	
Cost	Low		Medium			High	
Color	Black	Gray	White	Blue	Green	Red	Yellow

Worksheet 3: Property Values

Step 4: Determine Matching Material Families

On the MaterialTool Title page click on the button called "Materials." This takes you to the Materials Section. Click on the large button called "Properties." This takes you to the Materials Properties page. Notice the list of material families listed on the left: Ceramic, Cement, Fiber, and so on. In the right pane you see a list of properties such as Hardness, Impact Strength, Rigidity, and so on. Click on the values Low, Medium, or High to determine Material Families that match the choices from Worksheet 3: Property Values. List the matching families for Part A in Worksheet 4a, the matching families for Part B in Worksheet 4b, and the matching families for Part C in Worksheet 4c.

Part A				
Properties	**Value Selected**			**Matching Materials**
Hardness	Low	Medium	High	
Impact Strength	Low	Medium	High	
Rigidity	Low	Medium	High	
Abrasion Resistance	Low	Medium	High	
Optical Clarity	Low	Medium	High	
Chemical Resistance	Low	Medium	High	
Food Compatibility	Low	Medium	High	
Ultraviolet Stability	Low	Medium	High	
Moisture Resistance	Low	Medium	High	
Lubricity	Low	Medium	High	
Electrical Insulation	Low	Medium	High	
Heat Resistance	Low	Medium	High	
Recyclability	Low	Medium	High	
Degradability	Low	Medium	High	
Cost	Low	Medium	High	
Color	Black				
	Gray				
	White				
	Blue				
	Green				
	Red				
	Yellow				

Worksheet 4a: Matching Material Families

Step 5: Identify the Most Appropriate Material Families

List up to three material families that satisfy all the requirements listed in Worksheet 4a.

Part A					
1.		2.		3.	

Worksheet 5a: List of Matching Material Families

Part B				
Properties	**Value Selected**			**Matching Materials**
Hardness	Low	Medium	High	
Impact Strength	Low	Medium	High	
Rigidity	Low	Medium	High	
Abrasion Resistance	Low	Medium	High	
Optical Clarity	Low	Medium	High	
Chemical Resistance	Low	Medium	High	
Food Compatibility	Low	Medium	High	
Ultraviolet Stability	Low	Medium	High	
Moisture Resistance	Low	Medium	High	
Lubricity	Low	Medium	High	
Electrical Insulation	Low	Medium	High	
Heat Resistance	Low	Medium	High	
Recyclability	Low	Medium	High	
Degradability	Low	Medium	High	
Cost	Low	Medium	High	
Color	Black			
	Gray			
	White			
	Blue			
	Green			
	Red			
	Yellow			

Worksheet 4b: Matching Material Families

List up to three material families that satisfy all the requirements listed in Worksheet 4b.

Part B					
1.		2.		3.	

Worksheet 5b: List of Matching Material Families

Part C				
Properties	**Value Selected**			**Matching Materials**
Hardness	Low	Medium	High	
Impact Strength	Low	Medium	High	
Rigidity	Low	Medium	High	
Abrasion Resistance	Low	Medium	High	
Optical Clarity	Low	Medium	High	
Chemical Resistance	Low	Medium	High	
Food Compatibility	Low	Medium	High	
Ultraviolet Stability	Low	Medium	High	
Moisture Resistance	Low	Medium	High	
Lubricity	Low	Medium	High	
Electrical Insulation	Low	Medium	High	
Heat Resistance	Low	Medium	High	
Recyclability	Low	Medium	High	
Degradability	Low	Medium	High	
Cost	Low	Medium	High	
Color	Black			
	Gray			
	White			
	Blue			
	Green			
	Red			
	Yellow			

Worksheet 4c: Matching Material Families

List up to three material families that satisfy all the requirements listed in Worksheet 4c.

Part C					
1.		2.		3.	

Worksheet 5c: List of Matching Material Families

Step 6: Determine Matching Materials

Once again, go to the Materials Section and click on the large button called "Properties." This takes you to the main Materials Properties page. Referring to the list you created in Worksheet 3, click the arrow next to the first family. This brings up detailed materials. In the right pane, you see a list of properties such as Hardness, Impact Strength, Rigidity, and so on. Once again select the materials that match the choices in Worksheet 1. List these below in Worksheet 6: Matching Materials. List the matching materials for Part A in Worksheet 6a, the matching materials for Part B in Worksheet 6b, and the matching materials for Part C in Worksheet 6c.

Part A				
Properties	**Value Selected**			**Matching Materials**
Hardness	Low	Medium	High	
Impact Strength	Low	Medium	High	
Rigidity	Low	Medium	High	
Abrasion Resistance	Low	Medium	High	
Optical Clarity	Low	Medium	High	
Chemical Resistance	Low	Medium	High	
Food Compatibility	Low	Medium	High	
Ultraviolet Stability	Low	Medium	High	
Moisture Resistance	Low	Medium	High	
Lubricity	Low	Medium	High	
Electrical Insulation	Low	Medium	High	
Heat Resistance	Low	Medium	High	
Recyclability	Low	Medium	High	
Degradability	Low	Medium	High	
Cost	Low	Medium	High	
Color	Black			
	Gray			
	White			
	Blue			
	Green			
	Red			
	Yellow			

Worksheet 6a: Matching Materials

List up to three materials that best satisfy all the requirements listed in Worksheet 3.

Part A		
1.	2.	3.

Worksheet 7a: List of Matching Materials

Part B				
Properties	**Value Selected**			**Matching Materials**
Hardness	Low	Medium	High	
Impact Strength	Low	Medium	High	
Rigidity	Low	Medium	High	
Abrasion Resistance	Low	Medium	High	
Optical Clarity	Low	Medium	High	
Chemical Resistance	Low	Medium	High	
Food Compatibility	Low	Medium	High	
Ultraviolet Stability	Low	Medium	High	
Moisture Resistance	Low	Medium	High	
Lubricity	Low	Medium	High	
Electrical Insulation	Low	Medium	High	
Heat Resistance	Low	Medium	High	
Recyclability	Low	Medium	High	
Degradability	Low	Medium	High	
Cost	Low	Medium	High	
Color	Black			
	Gray			
	White			
	Blue			
	Green			
	Red			
	Yellow			

Worksheet 6b: Matching Materials

List up to three materials that best satisfy all the requirements listed in Worksheet 3.

Part B					
1.		2.		3.	

Worksheet 7b: List of Matching Materials

Part C				
Properties	**Value Selected**			**Matching Materials**
Hardness	Low	Medium	High	
Impact Strength	Low	Medium	High	
Rigidity	Low	Medium	High	
Abrasion Resistance	Low	Medium	High	
Optical Clarity	Low	Medium	High	
Chemical Resistance	Low	Medium	High	
Food Compatibility	Low	Medium	High	
Ultraviolet Stability	Low	Medium	High	
Moisture Resistance	Low	Medium	High	
Lubricity	Low	Medium	High	
Electrical Insulation	Low	Medium	High	
Heat Resistance	Low	Medium	High	
Recyclability	Low	Medium	High	
Degradability	Low	Medium	High	
Cost	Low	Medium	High	
Color	Black			
	Gray			
	White			
	Blue			
	Green			
	Red			
	Yellow			

Worksheet 6c: Matching Materials

List up to three materials that best satisfy all the requirements listed in Worksheet 3.

Part C		
1.	2.	3.

Worksheet 7c: List of Matching Materials

SELECTING PROCESSES

Step 8: Define the Form Requirements
Study the problem description, define the form characteristics desired of the product, and list these below in Worksheet 8: List of Form Requirements.

Part A
Part B
Part C

Worksheet 8: List of Form Requirements

Step 9: Determine the Stages by which the Desired Form is Achieved
Determine how the form requirements, described in Worksheet 8: List of Form Requirements, can be achieved as a series of successive stages. List each part and the necessary operations required to achieve the desired form in stages. If there are multiple parts in the product, call them out as well. List this information below in Worksheet 9: Stages of Form Generation.

Stage	What form is achieved in this stage?
	Part A
	Part B
	Part C

Worksheet 9: Stages of Form Generation

Step 10: Determine the Manufacturing Processes that are Relevant

Refer to the materials selections listed previously in Worksheet 7: List of Matching Materials, and the form requirements listed in Worksheet 9: Stages of Form Generation. Apply the procedure described below to determine the processes that are appropriate for each stage of processing each of the selected materials.

- Load MaterialTool and on the Title page click on the button called "**Processes.**" This takes you to the Processes section of MaterialTool.
- Click on any of the process groups buttons: "**Finishing**", "**Forming**", "**Joining**" or "**Machining**."
  This takes you to the selected processes group page. Notice that all major material families (such as Ceramic, Leather, and Wood) are listed in the definition. By clicking on these words, it is possible to determine the processes applicable to each material group.
- Click on the button with the name of an individual process. This takes you to that particular process page. By reading the process definition, design tips, and visual consequences, determine if the process can be used to achieve the desired form in the particular material.
- For a deeper understanding of the process, click on the green button named "**Show principle**."
  List the selected processes below in Worksheet 10: List of Selected Manufacturing Processes.

Step	Process	Description
Part A		
1.		
2.		
3.		
4.		
5.		
Part B		
1.		
2.		
3.		
4.		
5.		
Part C		
1.		
2.		
3.		
4.		
5.		

Worksheet 10: List of Selected Manufacturing Processes

FINAL SELECTION OF MATERIALS & PROCESSES

Step 11: List the Final Selections of Materials and Processes

Refer to the materials selections listed previously in Worksheet 7: List of Matching Materials, and the form requirements listed in Worksheet 10: List of Selected Manufacturing Processes. List the final choices of materials and manufacturing processes below in the Final Report.

FINAL REPORT	
Part A	
Material Selected	
Properties	
Rationale	
Processes Selected	
Operation	
Rationale	
Part B	
Material Selected	
Properties	
Rationale	
Processes Selected	
Operation	
Rationale	
Part C	
Material Selected	
Properties	
Rationale	
Processes Selected	
Operation	
Rationale	

Worksheet 11: Final Report

Unit 15: Coffee Pot

Problem
The picture shows a coffee pot. It is an inexpensive, mass-produced product. If you were to produce a similar product and were concerned about cost-effectiveness and ease of assembly, what specific materials and technologies would you select? Describe all steps in the production process.

SELECTING MATERIALS

Step 1: Identify the Parts of the Product
The product shown above may be made up of more than one part. Examine the photograph of the product, and identify all the parts of the product. List these in Worksheet 1: List of all Product Parts.

1.	
2.	
3.	
4.	
5.	
6.	
7.	
8.	
9.	
10.	

Worksheet 1: List of all Product Parts

Step 2: List Unique Parts
Some of the parts listed above may be identical parts. Identify the unique parts of the product. Consider mirrored parts, such as a left half and a right half of a product, to be identical parts. List up to three unique parts in Worksheet 2: Unique Product Parts.

Part A	
Part B	
Part C	

Worksheet 2: Unique Product Parts

Step 3: Define the Desired Property Values

First identify what performance characteristics are required for each part that has been identified above. Circle the property values for the material that satisfy the performance requirements for each part in Worksheet 3: Property Values below. Note that only some of the properties will be relevant for each part; leave the rest blank.

Part A							
Hardness	Low		Medium		High		
Impact Strength	Low		Medium		High		
Rigidity	Low		Medium		High		
Abrasion Resistance	Low		Medium		High		
Optical Clarity	Low		Medium		High		
Chemical Resistance	Low		Medium		High		
Food Compatibility	Low		Medium		High		
Ultraviolet Stability	Low		Medium		High		
Moisture Resistance	Low		Medium		High		
Lubricity	Low		Medium		High		
Electrical Insulation	Low		Medium		High		
Heat Resistance	Low		Medium		High		
Recyclability	Low		Medium		High		
Degradability	Low		Medium		High		
Cost	Low		Medium		High		
Color	Black	Gray	White	Blue	Green	Red	Yellow
Part B							
Hardness	Low		Medium		High		
Impact Strength	Low		Medium		High		
Rigidity	Low		Medium		High		
Abrasion Resistance	Low		Medium		High		
Optical Clarity	Low		Medium		High		
Chemical Resistance	Low		Medium		High		
Food Compatibility	Low		Medium		High		
Ultraviolet Stability	Low		Medium		High		
Moisture Resistance	Low		Medium		High		
Lubricity	Low		Medium		High		
Electrical Insulation	Low		Medium		High		
Heat Resistance	Low		Medium		High		
Recyclability	Low		Medium		High		
Degradability	Low		Medium		High		
Cost	Low		Medium		High		
Color	Black	Gray	White	Blue	Green	Red	Yellow
Part C							
Hardness	Low		Medium		High		
Impact Strength	Low		Medium		High		
Rigidity	Low		Medium		High		
Abrasion Resistance	Low		Medium		High		
Optical Clarity	Low		Medium		High		
Chemical Resistance	Low		Medium		High		
Food Compatibility	Low		Medium		High		
Ultraviolet Stability	Low		Medium		High		
Moisture Resistance	Low		Medium		High		
Lubricity	Low		Medium		High		
Electrical Insulation	Low		Medium		High		
Heat Resistance	Low		Medium		High		
Recyclability	Low		Medium		High		
Degradability	Low		Medium		High		
Cost	Low		Medium		High		
Color	Black	Gray	White	Blue	Green	Red	Yellow

Worksheet 3: Property Values

Step 4: Determine Matching Material Families

On the MaterialTool Title page click on the button called "Materials." This takes you to the Materials Section. Click on the large button called "Properties." This takes you to the Materials Properties page. Notice the list of material families listed on the left: Ceramic, Cement, Fiber, and so on. In the right pane you see a list of properties such as Hardness, Impact Strength, Rigidity, and so on. Click on the values Low, Medium, or High to determine Material Families that match the choices from Worksheet 3: Property Values. List the matching families for Part A in Worksheet 4a, the matching families for Part B in Worksheet 4b, and the matching families for Part C in Worksheet 4c.

Part A				
Properties	**Value Selected**			**Matching Materials**
Hardness	Low	Medium	High	
Impact Strength	Low	Medium	High	
Rigidity	Low	Medium	High	
Abrasion Resistance	Low	Medium	High	
Optical Clarity	Low	Medium	High	
Chemical Resistance	Low	Medium	High	
Food Compatibility	Low	Medium	High	
Ultraviolet Stability	Low	Medium	High	
Moisture Resistance	Low	Medium	High	
Lubricity	Low	Medium	High	
Electrical Insulation	Low	Medium	High	
Heat Resistance	Low	Medium	High	
Recyclability	Low	Medium	High	
Degradability	Low	Medium	High	
Cost	Low	Medium	High	
Color	Black			
	Gray			
	White			
	Blue			
	Green			
	Red			
	Yellow			

Worksheet 4a: Matching Material Families

Step 5: Identify the Most Appropriate Material Families

List up to three material families that satisfy all the requirements listed in Worksheet 4a.

Part A		
1.	2.	3.

Worksheet 5a: List of Matching Material Families

Part B				
Properties	**Value Selected**			**Matching Materials**
Hardness	Low	Medium	High	
Impact Strength	Low	Medium	High	
Rigidity	Low	Medium	High	
Abrasion Resistance	Low	Medium	High	
Optical Clarity	Low	Medium	High	
Chemical Resistance	Low	Medium	High	
Food Compatibility	Low	Medium	High	
Ultraviolet Stability	Low	Medium	High	
Moisture Resistance	Low	Medium	High	
Lubricity	Low	Medium	High	
Electrical Insulation	Low	Medium	High	
Heat Resistance	Low	Medium	High	
Recyclability	Low	Medium	High	
Degradability	Low	Medium	High	
Cost	Low	Medium	High	
Color	Black			
	Gray			
	White			
	Blue			
	Green			
	Red			
	Yellow			

Worksheet 4b: Matching Material Families

List up to three material families that satisfy all the requirements listed in Worksheet 4b.

Part B		
1.	2.	3.

Worksheet 5b: List of Matching Material Families

Part C				
Properties	**Value Selected**			**Matching Materials**
Hardness	Low	Medium	High	
Impact Strength	Low	Medium	High	
Rigidity	Low	Medium	High	
Abrasion Resistance	Low	Medium	High	
Optical Clarity	Low	Medium	High	
Chemical Resistance	Low	Medium	High	
Food Compatibility	Low	Medium	High	
Ultraviolet Stability	Low	Medium	High	
Moisture Resistance	Low	Medium	High	
Lubricity	Low	Medium	High	
Electrical Insulation	Low	Medium	High	
Heat Resistance	Low	Medium	High	
Recyclability	Low	Medium	High	
Degradability	Low	Medium	High	
Cost	Low	Medium	High	
Color	Black			
	Gray			
	White			
	Blue			
	Green			
	Red			
	Yellow			

Worksheet 4c: Matching Material Families

List up to three material families that satisfy all the requirements listed in Worksheet 4c.

Part C		
1.	2.	3.

Worksheet 5c: List of Matching Material Families

Step 6: Determine Matching Materials
Once again, go to the Materials Section and click on the large button called "Properties." This takes you to the main Materials Properties page. Referring to the list you created in Worksheet 3, click the arrow next to the first family. This brings up detailed materials. In the right pane, you see a list of properties such as Hardness, Impact Strength, Rigidity, and so on. Once again select the materials that match the choices in Worksheet 1. List these below in Worksheet 6: Matching Materials. List the matching materials for Part A in Worksheet 6a, the matching materials for Part B in Worksheet 6b, and the matching materials for Part C in Worksheet 6c.

Part A				
Properties	**Value Selected**			**Matching Materials**
Hardness	Low	Medium	High	
Impact Strength	Low	Medium	High	
Rigidity	Low	Medium	High	
Abrasion Resistance	Low	Medium	High	
Optical Clarity	Low	Medium	High	
Chemical Resistance	Low	Medium	High	
Food Compatibility	Low	Medium	High	
Ultraviolet Stability	Low	Medium	High	
Moisture Resistance	Low	Medium	High	
Lubricity	Low	Medium	High	
Electrical Insulation	Low	Medium	High	
Heat Resistance	Low	Medium	High	
Recyclability	Low	Medium	High	
Degradability	Low	Medium	High	
Cost	Low	Medium	High	
Color	Black			
	Gray			
	White			
	Blue			
	Green			
	Red			
	Yellow			

Worksheet 6a: Matching Materials

List up to three materials that best satisfy all the requirements listed in Worksheet 3.

Part A		
1.	2.	3.

Worksheet 7a: List of Matching Materials

Part B				
Properties	**Value Selected**			**Matching Materials**
Hardness	Low	Medium	High	
Impact Strength	Low	Medium	High	
Rigidity	Low	Medium	High	
Abrasion Resistance	Low	Medium	High	
Optical Clarity	Low	Medium	High	
Chemical Resistance	Low	Medium	High	
Food Compatibility	Low	Medium	High	
Ultraviolet Stability	Low	Medium	High	
Moisture Resistance	Low	Medium	High	
Lubricity	Low	Medium	High	
Electrical Insulation	Low	Medium	High	
Heat Resistance	Low	Medium	High	
Recyclability	Low	Medium	High	
Degradability	Low	Medium	High	
Cost	Low	Medium	High	
Color	Black			
	Gray			
	White			
	Blue			
	Green			
	Red			
	Yellow			

Worksheet 6b: Matching Materials

List up to three materials that best satisfy all the requirements listed in Worksheet 3.

Part B					
1.		2.		3.	

Worksheet 7b: List of Matching Materials

Part C				
Properties	**Value Selected**			**Matching Materials**
Hardness	Low	Medium	High	
Impact Strength	Low	Medium	High	
Rigidity	Low	Medium	High	
Abrasion Resistance	Low	Medium	High	
Optical Clarity	Low	Medium	High	
Chemical Resistance	Low	Medium	High	
Food Compatibility	Low	Medium	High	
Ultraviolet Stability	Low	Medium	High	
Moisture Resistance	Low	Medium	High	
Lubricity	Low	Medium	High	
Electrical Insulation	Low	Medium	High	
Heat Resistance	Low	Medium	High	
Recyclability	Low	Medium	High	
Degradability	Low	Medium	High	
Cost	Low	Medium	High	
Color	Black			
	Gray			
	White			
	Blue			
	Green			
	Red			
	Yellow			

Worksheet 6c: Matching Materials

List up to three materials that best satisfy all the requirements listed in Worksheet 3.

Part C		
1.	2.	3.

Worksheet 7c: List of Matching Materials

SELECTING PROCESSES

Step 8: Define the Form Requirements
Study the problem description, define the form characteristics desired of the product, and list these below in Worksheet 8: List of Form Requirements.

Part A
Part B
Part C

Worksheet 8: List of Form Requirements

Step 9: Determine the Stages by which the Desired Form is Achieved
Determine how the form requirements, described in Worksheet 8: List of Form Requirements, can be achieved as a series of successive stages. List each part and the necessary operations required to achieve the desired form in stages. If there are multiple parts in the product, call them out as well. List this information below in Worksheet 9: Stages of Form Generation.

Stage	What form is achieved in this stage?
Part A	
Part B	
Part C	

Worksheet 9: Stages of Form Generation

Step 10: Determine the Manufacturing Processes that are Relevant
Refer to the materials selections listed previously in Worksheet 7: List of Matching Materials, and the form requirements listed in Worksheet 9: Stages of Form Generation. Apply the procedure described below to determine the processes that are appropriate for each stage of processing each of the selected materials.

- Load MaterialTool and on the Title page click on the button called "**Processes.**" This takes you to the Processes section of MaterialTool.
- Click on any of the process groups buttons: "**Finishing**", "**Forming**", "**Joining**" or "**Machining**."
  This takes you to the selected processes group page. Notice that all major material families (such as Ceramic, Leather, and Wood) are listed in the definition. By clicking on these words, it is possible to determine the processes applicable to each material group.
- Click on the button with the name of an individual process. This takes you to that particular process page. By reading the process definition, design tips, and visual consequences, determine if the process can be used to achieve the desired form in the particular material.
- For a deeper understanding of the process, click on the green button named "**Show principle**."
  List the selected processes below in Worksheet 10: List of Selected Manufacturing Processes.

Step	Process	Description
Part A		
1.		
2.		
3.		
4.		
5.		
Part B		
1.		
2.		
3.		
4.		
5.		
Part C		
1.		
2.		
3.		
4.		
5.		

Worksheet 10: List of Selected Manufacturing Processes

FINAL SELECTION OF MATERIALS & PROCESSES

Step 11: List the Final Selections of Materials and Processes
Refer to the materials selections listed previously in Worksheet 7: List of Matching Materials, and the form requirements listed in Worksheet 10: List of Selected Manufacturing Processes. List the final choices of materials and manufacturing processes below in the Final Report.

\	FINAL REPORT
	Part A
Material Selected	
Properties	
Rationale	
Processes Selected	
Operation	
Rationale	
	Part B
Material Selected	
Properties	
Rationale	
Processes Selected	
Operation	
Rationale	
	Part C
Material Selected	
Properties	
Rationale	
Processes Selected	
Operation	
Rationale	

Worksheet 11: Final Report

Unit 16: Clothes Clip

Problem
The picture shows a clothes clip. If you were to produce similar clips and were concerned about cost-effectiveness and ease of assembly, what specific materials and technologies would you select? Describe all steps in the production process.

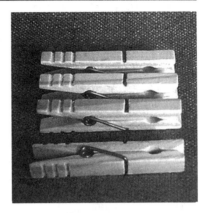

SELECTING MATERIALS

Step 1: Identify the Parts of the Product
The product shown above may be made up of more than one part. Examine the photograph of the product, and identify all the parts of the product. List these in Worksheet 1: List of all Product Parts.

1.	
2.	
3.	
4.	
5.	
6.	
7.	
8.	
9.	
10.	

Worksheet 1: List of all Product Parts

Step 2: List Unique Parts
Some of the parts listed above may be identical parts. Identify the unique parts of the product. Consider mirrored parts, such as a left half and a right half of a product, to be identical parts. List up to three unique parts in Worksheet 2: Unique Product Parts.

Part A	
Part B	
Part C	

Worksheet 2: Unique Product Parts

179

Step 3: Define the Desired Property Values

First identify what performance characteristics are required for each part that has been identified above. Circle the property values for the material that satisfy the performance requirements for each part in Worksheet 3: Property Values below. Note that only some of the properties will be relevant for each part; leave the rest blank.

Part A							
Hardness	Low		Medium			High	
Impact Strength	Low		Medium			High	
Rigidity	Low		Medium			High	
Abrasion Resistance	Low		Medium			High	
Optical Clarity	Low		Medium			High	
Chemical Resistance	Low		Medium			High	
Food Compatibility	Low		Medium			High	
Ultraviolet Stability	Low		Medium			High	
Moisture Resistance	Low		Medium			High	
Lubricity	Low		Medium			High	
Electrical Insulation	Low		Medium			High	
Heat Resistance	Low		Medium			High	
Recyclability	Low		Medium			High	
Degradability	Low		Medium			High	
Cost	Low		Medium			High	
Color	Black	Gray	White	Blue	Green	Red	Yellow
Part B							
Hardness	Low		Medium			High	
Impact Strength	Low		Medium			High	
Rigidity	Low		Medium			High	
Abrasion Resistance	Low		Medium			High	
Optical Clarity	Low		Medium			High	
Chemical Resistance	Low		Medium			High	
Food Compatibility	Low		Medium			High	
Ultraviolet Stability	Low		Medium			High	
Moisture Resistance	Low		Medium			High	
Lubricity	Low		Medium			High	
Electrical Insulation	Low		Medium			High	
Heat Resistance	Low		Medium			High	
Recyclability	Low		Medium			High	
Degradability	Low		Medium			High	
Cost	Low		Medium			High	
Color	Black	Gray	White	Blue	Green	Red	Yellow
Part C							
Hardness	Low		Medium			High	
Impact Strength	Low		Medium			High	
Rigidity	Low		Medium			High	
Abrasion Resistance	Low		Medium			High	
Optical Clarity	Low		Medium			High	
Chemical Resistance	Low		Medium			High	
Food Compatibility	Low		Medium			High	
Ultraviolet Stability	Low		Medium			High	
Moisture Resistance	Low		Medium			High	
Lubricity	Low		Medium			High	
Electrical Insulation	Low		Medium			High	
Heat Resistance	Low		Medium			High	
Recyclability	Low		Medium			High	
Degradability	Low		Medium			High	
Cost	Low		Medium			High	
Color	Black	Gray	White	Blue	Green	Red	Yellow

Worksheet 3: Property Values

Step 4: Determine Matching Material Families

On the MaterialTool Title page click on the button called "Materials." This takes you to the Materials Section. Click on the large button called "Properties." This takes you to the Materials Properties page. Notice the list of material families listed on the left: Ceramic, Cement, Fiber, and so on. In the right pane you see a list of properties such as Hardness, Impact Strength, Rigidity, and so on. Click on the values Low, Medium, or High to determine Material Families that match the choices from Worksheet 3: Property Values. List the matching families for Part A in Worksheet 4a, the matching families for Part B in Worksheet 4b, and the matching families for Part C in Worksheet 4c.

Part A				
Properties	**Value Selected**			**Matching Materials**
Hardness	Low	Medium	High	
Impact Strength	Low	Medium	High	
Rigidity	Low	Medium	High	
Abrasion Resistance	Low	Medium	High	
Optical Clarity	Low	Medium	High	
Chemical Resistance	Low	Medium	High	
Food Compatibility	Low	Medium	High	
Ultraviolet Stability	Low	Medium	High	
Moisture Resistance	Low	Medium	High	
Lubricity	Low	Medium	High	
Electrical Insulation	Low	Medium	High	
Heat Resistance	Low	Medium	High	
Recyclability	Low	Medium	High	
Degradability	Low	Medium	High	
Cost	Low	Medium	High	
Color	Black			
	Gray			
	White			
	Blue			
	Green			
	Red			
	Yellow			

Worksheet 4a: Matching Material Families

Step 5: Identify the Most Appropriate Material Families

List up to three material families that satisfy all the requirements listed in Worksheet 4a.

Part A					
1.		2.		3.	

Worksheet 5a: List of Matching Material Families

Part B				
Properties	**Value Selected**			**Matching Materials**
Hardness	Low	Medium	High	
Impact Strength	Low	Medium	High	
Rigidity	Low	Medium	High	
Abrasion Resistance	Low	Medium	High	
Optical Clarity	Low	Medium	High	
Chemical Resistance	Low	Medium	High	
Food Compatibility	Low	Medium	High	
Ultraviolet Stability	Low	Medium	High	
Moisture Resistance	Low	Medium	High	
Lubricity	Low	Medium	High	
Electrical Insulation	Low	Medium	High	
Heat Resistance	Low	Medium	High	
Recyclability	Low	Medium	High	
Degradability	Low	Medium	High	
Cost	Low	Medium	High	
Color	Black			
	Gray			
	White			
	Blue			
	Green			
	Red			
	Yellow			

Worksheet 4b: Matching Material Families

List up to three material families that satisfy all the requirements listed in Worksheet 4b.

Part B		
1.	2.	3.

Worksheet 5b: List of Matching Material Families

Part C				
Properties	**Value Selected**			**Matching Materials**
Hardness	Low	Medium	High	
Impact Strength	Low	Medium	High	
Rigidity	Low	Medium	High	
Abrasion Resistance	Low	Medium	High	
Optical Clarity	Low	Medium	High	
Chemical Resistance	Low	Medium	High	
Food Compatibility	Low	Medium	High	
Ultraviolet Stability	Low	Medium	High	
Moisture Resistance	Low	Medium	High	
Lubricity	Low	Medium	High	
Electrical Insulation	Low	Medium	High	
Heat Resistance	Low	Medium	High	
Recyclability	Low	Medium	High	
Degradability	Low	Medium	High	
Cost	Low	Medium	High	
Color	Black			
	Gray			
	White			
	Blue			
	Green			
	Red			
	Yellow			

Worksheet 4c: Matching Material Families

List up to three material families that satisfy all the requirements listed in Worksheet 4c.

Part C					
1.		2.		3.	

Worksheet 5c: List of Matching Material Families

Step 6: Determine Matching Materials

Once again, go to the Materials Section and click on the large button called "Properties." This takes you to the main Materials Properties page. Referring to the list you created in Worksheet 3, click the arrow next to the first family. This brings up detailed materials. In the right pane, you see a list of properties such as Hardness, Impact Strength, Rigidity, and so on. Once again select the materials that match the choices in Worksheet 1. List these below in Worksheet 6: Matching Materials. List the matching materials for Part A in Worksheet 6a, the matching materials for Part B in Worksheet 6b, and the matching materials for Part C in Worksheet 6c.

Part A				
Properties	**Value Selected**			**Matching Materials**
Hardness	Low	Medium	High	
Impact Strength	Low	Medium	High	
Rigidity	Low	Medium	High	
Abrasion Resistance	Low	Medium	High	
Optical Clarity	Low	Medium	High	
Chemical Resistance	Low	Medium	High	
Food Compatibility	Low	Medium	High	
Ultraviolet Stability	Low	Medium	High	
Moisture Resistance	Low	Medium	High	
Lubricity	Low	Medium	High	
Electrical Insulation	Low	Medium	High	
Heat Resistance	Low	Medium	High	
Recyclability	Low	Medium	High	
Degradability	Low	Medium	High	
Cost	Low	Medium	High	
Color	Black			
	Gray			
	White			
	Blue			
	Green			
	Red			
	Yellow			

Worksheet 6a: Matching Materials

List up to three materials that best satisfy all the requirements listed in Worksheet 3.

Part A		
1.	2.	3.

Worksheet 7a: List of Matching Materials

Part B				
Properties	**Value Selected**			**Matching Materials**
Hardness	Low	Medium	High	
Impact Strength	Low	Medium	High	
Rigidity	Low	Medium	High	
Abrasion Resistance	Low	Medium	High	
Optical Clarity	Low	Medium	High	
Chemical Resistance	Low	Medium	High	
Food Compatibility	Low	Medium	High	
Ultraviolet Stability	Low	Medium	High	
Moisture Resistance	Low	Medium	High	
Lubricity	Low	Medium	High	
Electrical Insulation	Low	Medium	High	
Heat Resistance	Low	Medium	High	
Recyclability	Low	Medium	High	
Degradability	Low	Medium	High	
Cost	Low	Medium	High	
Color	Black			
	Gray			
	White			
	Blue			
	Green			
	Red			
	Yellow			

Worksheet 6b: Matching Materials

List up to three materials that best satisfy all the requirements listed in Worksheet 3.

Part B					
1.		2.		3.	

Worksheet 7b: List of Matching Materials

Part C				
Properties	**Value Selected**			**Matching Materials**
Hardness	Low	Medium	High	
Impact Strength	Low	Medium	High	
Rigidity	Low	Medium	High	
Abrasion Resistance	Low	Medium	High	
Optical Clarity	Low	Medium	High	
Chemical Resistance	Low	Medium	High	
Food Compatibility	Low	Medium	High	
Ultraviolet Stability	Low	Medium	High	
Moisture Resistance	Low	Medium	High	
Lubricity	Low	Medium	High	
Electrical Insulation	Low	Medium	High	
Heat Resistance	Low	Medium	High	
Recyclability	Low	Medium	High	
Degradability	Low	Medium	High	
Cost	Low	Medium	High	
Color	Black			
	Gray			
	White			
	Blue			
	Green			
	Red			
	Yellow			

Worksheet 6c: Matching Materials

List up to three materials that best satisfy all the requirements listed in Worksheet 3.

Part C		
1.	2.	3.

Worksheet 7c: List of Matching Materials

SELECTING PROCESSES

Step 8: Define the Form Requirements
Study the problem description, define the form characteristics desired of the product, and list these below in Worksheet 8: List of Form Requirements.

Part A
Part B
Part C

Worksheet 8: List of Form Requirements

Step 9: Determine the Stages by which the Desired Form is Achieved
Determine how the form requirements, described in Worksheet 8: List of Form Requirements, can be achieved as a series of successive stages. List each part and the necessary operations required to achieve the desired form in stages. If there are multiple parts in the product, call them out as well. List this information below in Worksheet 9: Stages of Form Generation.

Stage	What form is achieved in this stage?
	Part A
	Part B
	Part C

Worksheet 9: Stages of Form Generation

Step 10: Determine the Manufacturing Processes that are Relevant

Refer to the materials selections listed previously in Worksheet 7: List of Matching Materials, and the form requirements listed in Worksheet 9: Stages of Form Generation. Apply the procedure described below to determine the processes that are appropriate for each stage of processing each of the selected materials.

- Load MaterialTool and on the Title page click on the button called "**Processes.**" This takes you to the Processes section of MaterialTool.
- Click on any of the process groups buttons: "**Finishing**", "**Forming**", "**Joining**" or "**Machining**." This takes you to the selected processes group page. Notice that all major material families (such as Ceramic, Leather, and Wood) are listed in the definition. By clicking on these words, it is possible to determine the processes applicable to each material group.
- Click on the button with the name of an individual process. This takes you to that particular process page. By reading the process definition, design tips, and visual consequences, determine if the process can be used to achieve the desired form in the particular material.
- For a deeper understanding of the process, click on the green button named "**Show principle**."

List the selected processes below in Worksheet 10: List of Selected Manufacturing Processes.

Step	Process	Description
Part A		
1.		
2.		
3.		
4.		
5.		
Part B		
1.		
2.		
3.		
4.		
5.		
Part C		
1.		
2.		
3.		
4.		
5.		

Worksheet 10: List of Selected Manufacturing Processes

FINAL SELECTION OF MATERIALS & PROCESSES

Step 11: List the Final Selections of Materials and Processes
Refer to the materials selections listed previously in Worksheet 7: List of Matching Materials, and the form requirements listed in Worksheet 10: List of Selected Manufacturing Processes. List the final choices of materials and manufacturing processes below in the Final Report.

FINAL REPORT	
Part A	
Material Selected	
Properties	
Rationale	
Processes Selected	
Operation	
Rationale	
Part B	
Material Selected	
Properties	
Rationale	
Processes Selected	
Operation	
Rationale	
Part C	
Material Selected	
Properties	
Rationale	
Processes Selected	
Operation	
Rationale	

Worksheet 11: Final Report

Unit 17: Grater

Problem

The picture on the right shows a grater which is available at a reasonable price and consists of two parts: the blade and a handle. What materials are the two parts of the grater made of? What are the processes used for manufacturing the grater?

SELECTING MATERIALS

Step 1: Identify the Parts of the Product

The product shown above may be made up of more than one part. Examine the photograph of the product, and identify all the parts of the product. List these in Worksheet 1: List of all Product Parts.

1.	
2.	
3.	
4.	
5.	
6.	
7.	
8.	
9.	
10.	

Worksheet 1: List of all Product Parts

Step 2: List Unique Parts

Some of the parts listed above may be identical parts. Identify the unique parts of the product. Consider mirrored parts, such as a left half and a right half of a product, to be identical parts. List up to three unique parts in Worksheet 2: Unique Product Parts.

Part A	
Part B	
Part C	

Worksheet 2: Unique Product Parts

Step 3: Define the Desired Property Values
First identify what performance characteristics are required for each part that has been identified above. Circle the property values for the material that satisfy the performance requirements for each part in Worksheet 3: Property Values below. Note that only some of the properties will be relevant for each part; leave the rest blank.

Part A							
Hardness		Low		Medium		High	
Impact Strength		Low		Medium		High	
Rigidity		Low		Medium		High	
Abrasion Resistance		Low		Medium		High	
Optical Clarity		Low		Medium		High	
Chemical Resistance		Low		Medium		High	
Food Compatibility		Low		Medium		High	
Ultraviolet Stability		Low		Medium		High	
Moisture Resistance		Low		Medium		High	
Lubricity		Low		Medium		High	
Electrical Insulation		Low		Medium		High	
Heat Resistance		Low		Medium		High	
Recyclability		Low		Medium		High	
Degradability		Low		Medium		High	
Cost		Low		Medium		High	
Color	Black	Gray	White	Blue	Green	Red	Yellow
Part B							
Hardness		Low		Medium		High	
Impact Strength		Low		Medium		High	
Rigidity		Low		Medium		High	
Abrasion Resistance		Low		Medium		High	
Optical Clarity		Low		Medium		High	
Chemical Resistance		Low		Medium		High	
Food Compatibility		Low		Medium		High	
Ultraviolet Stability		Low		Medium		High	
Moisture Resistance		Low		Medium		High	
Lubricity		Low		Medium		High	
Electrical Insulation		Low		Medium		High	
Heat Resistance		Low		Medium		High	
Recyclability		Low		Medium		High	
Degradability		Low		Medium		High	
Cost		Low		Medium		High	
Color	Black	Gray	White	Blue	Green	Red	Yellow
Part C							
Hardness		Low		Medium		High	
Impact Strength		Low		Medium		High	
Rigidity		Low		Medium		High	
Abrasion Resistance		Low		Medium		High	
Optical Clarity		Low		Medium		High	
Chemical Resistance		Low		Medium		High	
Food Compatibility		Low		Medium		High	
Ultraviolet Stability		Low		Medium		High	
Moisture Resistance		Low		Medium		High	
Lubricity		Low		Medium		High	
Electrical Insulation		Low		Medium		High	
Heat Resistance		Low		Medium		High	
Recyclability		Low		Medium		High	
Degradability		Low		Medium		High	
Cost		Low		Medium		High	
Color	Black	Gray	White	Blue	Green	Red	Yellow

Worksheet 3: Property Values

Step 4: Determine Matching Material Families
On the MaterialTool Title page click on the button called "Materials." This takes you to the Materials Section. Click on the large button called "Properties." This takes you to the Materials Properties page. Notice the list of material families listed on the left: Ceramic, Cement, Fiber, and so on. In the right pane you see a list of properties such as Hardness, Impact Strength, Rigidity, and so on. Click on the values Low, Medium, or High to determine Material Families that match the choices from Worksheet 3: Property Values. List the matching families for Part A in Worksheet 4a, the matching families for Part B in Worksheet 4b, and the matching families for Part C in Worksheet 4c.

Part A				
Properties	**Value Selected**			**Matching Materials**
Hardness	Low	Medium	High	
Impact Strength	Low	Medium	High	
Rigidity	Low	Medium	High	
Abrasion Resistance	Low	Medium	High	
Optical Clarity	Low	Medium	High	
Chemical Resistance	Low	Medium	High	
Food Compatibility	Low	Medium	High	
Ultraviolet Stability	Low	Medium	High	
Moisture Resistance	Low	Medium	High	
Lubricity	Low	Medium	High	
Electrical Insulation	Low	Medium	High	
Heat Resistance	Low	Medium	High	
Recyclability	Low	Medium	High	
Degradability	Low	Medium	High	
Cost	Low	Medium	High	
Color	Black			
	Gray			
	White			
	Blue			
	Green			
	Red			
	Yellow			

Worksheet 4a: Matching Material Families

Step 5: Identify the Most Appropriate Material Families
List up to three material families that satisfy all the requirements listed in Worksheet 4a.

Part A					
1.		2.		3.	

Worksheet 5a: List of Matching Material Families

Part B				
Properties	**Value Selected**			**Matching Materials**
Hardness	Low	Medium	High	
Impact Strength	Low	Medium	High	
Rigidity	Low	Medium	High	
Abrasion Resistance	Low	Medium	High	
Optical Clarity	Low	Medium	High	
Chemical Resistance	Low	Medium	High	
Food Compatibility	Low	Medium	High	
Ultraviolet Stability	Low	Medium	High	
Moisture Resistance	Low	Medium	High	
Lubricity	Low	Medium	High	
Electrical Insulation	Low	Medium	High	
Heat Resistance	Low	Medium	High	
Recyclability	Low	Medium	High	
Degradability	Low	Medium	High	
Cost	Low	Medium	High	
Color	Black			
	Gray			
	White			
	Blue			
	Green			
	Red			
	Yellow			

Worksheet 4b: Matching Material Families

List up to three material families that satisfy all the requirements listed in Worksheet 4b.

Part B		
1.	2.	3.

Worksheet 5b: List of Matching Material Families

Part C				
Properties	**Value Selected**			**Matching Materials**
Hardness	Low	Medium	High	
Impact Strength	Low	Medium	High	
Rigidity	Low	Medium	High	
Abrasion Resistance	Low	Medium	High	
Optical Clarity	Low	Medium	High	
Chemical Resistance	Low	Medium	High	
Food Compatibility	Low	Medium	High	
Ultraviolet Stability	Low	Medium	High	
Moisture Resistance	Low	Medium	High	
Lubricity	Low	Medium	High	
Electrical Insulation	Low	Medium	High	
Heat Resistance	Low	Medium	High	
Recyclability	Low	Medium	High	
Degradability	Low	Medium	High	
Cost	Low	Medium	High	
Color	Black			
	Gray			
	White			
	Blue			
	Green			
	Red			
	Yellow			

Worksheet 4c: Matching Material Families

List up to three material families that satisfy all the requirements listed in Worksheet 4c.

Part C		
1.	2.	3.

Worksheet 5c: List of Matching Material Families

Step 6: Determine Matching Materials
Once again, go to the Materials Section and click on the large button called "Properties." This takes you to the main Materials Properties page. Referring to the list you created in Worksheet 3, click the arrow next to the first family. This brings up detailed materials. In the right pane, you see a list of properties such as Hardness, Impact Strength, Rigidity, and so on. Once again select the materials that match the choices in Worksheet 1. List these below in Worksheet 6: Matching Materials. List the matching materials for Part A in Worksheet 6a, the matching materials for Part B in Worksheet 6b, and the matching materials for Part C in Worksheet 6c.

Part A				
Properties	**Value Selected**			**Matching Materials**
Hardness	Low	Medium	High	
Impact Strength	Low	Medium	High	
Rigidity	Low	Medium	High	
Abrasion Resistance	Low	Medium	High	
Optical Clarity	Low	Medium	High	
Chemical Resistance	Low	Medium	High	
Food Compatibility	Low	Medium	High	
Ultraviolet Stability	Low	Medium	High	
Moisture Resistance	Low	Medium	High	
Lubricity	Low	Medium	High	
Electrical Insulation	Low	Medium	High	
Heat Resistance	Low	Medium	High	
Recyclability	Low	Medium	High	
Degradability	Low	Medium	High	
Cost	Low	Medium	High	
Color	Black			
	Gray			
	White			
	Blue			
	Green			
	Red			
	Yellow			

Worksheet 6a: Matching Materials

List up to three materials that best satisfy all the requirements listed in Worksheet 3.

Part A		
1.	2.	3.

Worksheet 7a: List of Matching Materials

Part B				
Properties	**Value Selected**			**Matching Materials**
Hardness	Low	Medium	High	
Impact Strength	Low	Medium	High	
Rigidity	Low	Medium	High	
Abrasion Resistance	Low	Medium	High	
Optical Clarity	Low	Medium	High	
Chemical Resistance	Low	Medium	High	
Food Compatibility	Low	Medium	High	
Ultraviolet Stability	Low	Medium	High	
Moisture Resistance	Low	Medium	High	
Lubricity	Low	Medium	High	
Electrical Insulation	Low	Medium	High	
Heat Resistance	Low	Medium	High	
Recyclability	Low	Medium	High	
Degradability	Low	Medium	High	
Cost	Low	Medium	High	
Color	Black			
	Gray			
	White			
	Blue			
	Green			
	Red			
	Yellow			

Worksheet 6b: Matching Materials

List up to three materials that best satisfy all the requirements listed in Worksheet 3.

Part B		
1.	2.	3.

Worksheet 7b: List of Matching Materials

Part C				
Properties	**Value Selected**			**Matching Materials**
Hardness	Low	Medium	High	
Impact Strength	Low	Medium	High	
Rigidity	Low	Medium	High	
Abrasion Resistance	Low	Medium	High	
Optical Clarity	Low	Medium	High	
Chemical Resistance	Low	Medium	High	
Food Compatibility	Low	Medium	High	
Ultraviolet Stability	Low	Medium	High	
Moisture Resistance	Low	Medium	High	
Lubricity	Low	Medium	High	
Electrical Insulation	Low	Medium	High	
Heat Resistance	Low	Medium	High	
Recyclability	Low	Medium	High	
Degradability	Low	Medium	High	
Cost	Low	Medium	High	
Color	Black			
	Gray			
	White			
	Blue			
	Green			
	Red			
	Yellow			

Worksheet 6c: Matching Materials

List up to three materials that best satisfy all the requirements listed in Worksheet 3.

Part C				
1.		2.	3.	

Worksheet 7c: List of Matching Materials

SELECTING PROCESSES

Step 8: Define the Form Requirements
Study the problem description, define the form characteristics desired of the product, and list these below in Worksheet 8: List of Form Requirements.

Part A
Part B
Part C

Worksheet 8: List of Form Requirements

Step 9: Determine the Stages by which the Desired Form is Achieved
Determine how the form requirements, described in Worksheet 8: List of Form Requirements, can be achieved as a series of successive stages. List each part and the necessary operations required to achieve the desired form in stages. If there are multiple parts in the product, call them out as well. List this information below in Worksheet 9: Stages of Form Generation.

Stage	What form is achieved in this stage?
	Part A
	Part B
	Part C

Worksheet 9: Stages of Form Generation

Step 10: Determine the Manufacturing Processes that are Relevant
Refer to the materials selections listed previously in Worksheet 7: List of Matching Materials, and the form requirements listed in Worksheet 9: Stages of Form Generation. Apply the procedure described below to determine the processes that are appropriate for each stage of processing each of the selected materials.
- Load MaterialTool and on the Title page click on the button called "**Processes.**" This takes you to the Processes section of MaterialTool.
- Click on any of the process groups buttons: "**Finishing**", "**Forming**", "**Joining**" or "**Machining**."
  This takes you to the selected processes group page. Notice that all major material families (such as Ceramic, Leather, and Wood) are listed in the definition. By clicking on these words, it is possible to determine the processes applicable to each material group.
- Click on the button with the name of an individual process. This takes you to that particular process page. By reading the process definition, design tips, and visual consequences, determine if the process can be used to achieve the desired form in the particular material.
- For a deeper understanding of the process, click on the green button named "**Show principle**."
  List the selected processes below in Worksheet 10: List of Selected Manufacturing Processes.

Step	Process	Description
Part A		
1.		
2.		
3.		
4.		
5.		
Part B		
1.		
2.		
3.		
4.		
5.		
Part C		
1.		
2.		
3.		
4.		
5.		

Worksheet 10: List of Selected Manufacturing Processes

FINAL SELECTION OF MATERIALS & PROCESSES

Step 11: List the Final Selections of Materials and Processes

Refer to the materials selections listed previously in Worksheet 7: List of Matching Materials, and the form requirements listed in Worksheet 10: List of Selected Manufacturing Processes. List the final choices of materials and manufacturing processes below in the Final Report.

\	FINAL REPORT
\	**Part A**
Material Selected	
Properties	
Rationale	
Processes Selected	
Operation	
Rationale	
\	**Part B**
Material Selected	
Properties	
Rationale	
Processes Selected	
Operation	
Rationale	
\	**Part C**
Material Selected	
Properties	
Rationale	
Processes Selected	
Operation	
Rationale	

Worksheet 11: Final Report

Unit 18: Book

Problem
The picture shows a book that documents the winners of an annual design award. It has high-quality color reproductions of winning designs. Several thousand copies of the book were printed. What materials and processes are used to create this book?

SELECTING MATERIALS

Step 1: Identify the Parts of the Product
The product shown above may be made up of more than one part. Examine the photograph of the product, and identify all the parts of the product. List these in Worksheet 1: List of all Product Parts.

1.	
2.	
3.	
4.	
5.	
6.	
7.	
8.	
9.	
10.	

Worksheet 1: List of all Product Parts

Step 2: List Unique Parts
Some of the parts listed above may be identical parts. Identify the unique parts of the product. Consider mirrored parts, such as a left half and a right half of a product, to be identical parts. List up to three unique parts in Worksheet 2: Unique Product Parts.

Part A	
Part B	
Part C	

Worksheet 2: Unique Product Parts

Step 3: Define the Desired Property Values

First identify what performance characteristics are required for each part that has been identified above. Circle the property values for the material that satisfy the performance requirements for each part in Worksheet 3: Property Values below. Note that only some of the properties will be relevant for each part; leave the rest blank.

Part A							
Hardness	Low		Medium		High		
Impact Strength	Low		Medium		High		
Rigidity	Low		Medium		High		
Abrasion Resistance	Low		Medium		High		
Optical Clarity	Low		Medium		High		
Chemical Resistance	Low		Medium		High		
Food Compatibility	Low		Medium		High		
Ultraviolet Stability	Low		Medium		High		
Moisture Resistance	Low		Medium		High		
Lubricity	Low		Medium		High		
Electrical Insulation	Low		Medium		High		
Heat Resistance	Low		Medium		High		
Recyclability	Low		Medium		High		
Degradability	Low		Medium		High		
Cost	Low		Medium		High		
Color	Black	Gray	White	Blue	Green	Red	Yellow
Part B							
Hardness	Low		Medium		High		
Impact Strength	Low		Medium		High		
Rigidity	Low		Medium		High		
Abrasion Resistance	Low		Medium		High		
Optical Clarity	Low		Medium		High		
Chemical Resistance	Low		Medium		High		
Food Compatibility	Low		Medium		High		
Ultraviolet Stability	Low		Medium		High		
Moisture Resistance	Low		Medium		High		
Lubricity	Low		Medium		High		
Electrical Insulation	Low		Medium		High		
Heat Resistance	Low		Medium		High		
Recyclability	Low		Medium		High		
Degradability	Low		Medium		High		
Cost	Low		Medium		High		
Color	Black	Gray	White	Blue	Green	Red	Yellow
Part C							
Hardness	Low		Medium		High		
Impact Strength	Low		Medium		High		
Rigidity	Low		Medium		High		
Abrasion Resistance	Low		Medium		High		
Optical Clarity	Low		Medium		High		
Chemical Resistance	Low		Medium		High		
Food Compatibility	Low		Medium		High		
Ultraviolet Stability	Low		Medium		High		
Moisture Resistance	Low		Medium		High		
Lubricity	Low		Medium		High		
Electrical Insulation	Low		Medium		High		
Heat Resistance	Low		Medium		High		
Recyclability	Low		Medium		High		
Degradability	Low		Medium		High		
Cost	Low		Medium		High		
Color	Black	Gray	White	Blue	Green	Red	Yellow

Worksheet 3: Property Values

Step 4: Determine Matching Material Families
On the MaterialTool Title page click on the button called "Materials." This takes you to the Materials Section. Click on the large button called "Properties." This takes you to the Materials Properties page. Notice the list of material families listed on the left: Ceramic, Cement, Fiber, and so on. In the right pane you see a list of properties such as Hardness, Impact Strength, Rigidity, and so on. Click on the values Low, Medium, or High to determine Material Families that match the choices from Worksheet 3: Property Values. List the matching families for Part A in Worksheet 4a, the matching families for Part B in Worksheet 4b, and the matching families for Part C in Worksheet 4c.

Part A				
Properties	**Value Selected**			**Matching Materials**
Hardness	Low	Medium	High	
Impact Strength	Low	Medium	High	
Rigidity	Low	Medium	High	
Abrasion Resistance	Low	Medium	High	
Optical Clarity	Low	Medium	High	
Chemical Resistance	Low	Medium	High	
Food Compatibility	Low	Medium	High	
Ultraviolet Stability	Low	Medium	High	
Moisture Resistance	Low	Medium	High	
Lubricity	Low	Medium	High	
Electrical Insulation	Low	Medium	High	
Heat Resistance	Low	Medium	High	
Recyclability	Low	Medium	High	
Degradability	Low	Medium	High	
Cost	Low	Medium	High	
Color	Black			
	Gray			
	White			
	Blue			
	Green			
	Red			
	Yellow			

Worksheet 4a: Matching Material Families

Step 5: Identify the Most Appropriate Material Families
List up to three material families that satisfy all the requirements listed in Worksheet 4a.

Part A					
1.		2.		3.	

Worksheet 5a: List of Matching Material Families

Part B				
Properties	**Value Selected**			**Matching Materials**
Hardness	Low	Medium	High	
Impact Strength	Low	Medium	High	
Rigidity	Low	Medium	High	
Abrasion Resistance	Low	Medium	High	
Optical Clarity	Low	Medium	High	
Chemical Resistance	Low	Medium	High	
Food Compatibility	Low	Medium	High	
Ultraviolet Stability	Low	Medium	High	
Moisture Resistance	Low	Medium	High	
Lubricity	Low	Medium	High	
Electrical Insulation	Low	Medium	High	
Heat Resistance	Low	Medium	High	
Recyclability	Low	Medium	High	
Degradability	Low	Medium	High	
Cost	Low	Medium	High	
Color	Black			
	Gray			
	White			
	Blue			
	Green			
	Red			
	Yellow			

Worksheet 4b: Matching Material Families

List up to three material families that satisfy all the requirements listed in Worksheet 4b.

Part B					
1.		2.		3.	

Worksheet 5b: List of Matching Material Families

Part C				
Properties	**Value Selected**			**Matching Materials**
Hardness	Low	Medium	High	
Impact Strength	Low	Medium	High	
Rigidity	Low	Medium	High	
Abrasion Resistance	Low	Medium	High	
Optical Clarity	Low	Medium	High	
Chemical Resistance	Low	Medium	High	
Food Compatibility	Low	Medium	High	
Ultraviolet Stability	Low	Medium	High	
Moisture Resistance	Low	Medium	High	
Lubricity	Low	Medium	High	
Electrical Insulation	Low	Medium	High	
Heat Resistance	Low	Medium	High	
Recyclability	Low	Medium	High	
Degradability	Low	Medium	High	
Cost	Low	Medium	High	
Color		Black		
		Gray		
		White		
		Blue		
		Green		
		Red		
		Yellow		

Worksheet 4c: Matching Material Families

List up to three material families that satisfy all the requirements listed in Worksheet 4c.

Part C					
1.		2.		3.	

Worksheet 5c: List of Matching Material Families

Step 6: Determine Matching Materials

Once again, go to the Materials Section and click on the large button called "Properties." This takes you to the main Materials Properties page. Referring to the list you created in Worksheet 3, click the arrow next to the first family. This brings up detailed materials. In the right pane, you see a list of properties such as Hardness, Impact Strength, Rigidity, and so on. Once again select the materials that match the choices in Worksheet 1. List these below in Worksheet 6: Matching Materials. List the matching materials for Part A in Worksheet 6a, the matching materials for Part B in Worksheet 6b, and the matching materials for Part C in Worksheet 6c.

Part A				
Properties	**Value Selected**			**Matching Materials**
Hardness	Low	Medium	High	
Impact Strength	Low	Medium	High	
Rigidity	Low	Medium	High	
Abrasion Resistance	Low	Medium	High	
Optical Clarity	Low	Medium	High	
Chemical Resistance	Low	Medium	High	
Food Compatibility	Low	Medium	High	
Ultraviolet Stability	Low	Medium	High	
Moisture Resistance	Low	Medium	High	
Lubricity	Low	Medium	High	
Electrical Insulation	Low	Medium	High	
Heat Resistance	Low	Medium	High	
Recyclability	Low	Medium	High	
Degradability	Low	Medium	High	
Cost	Low	Medium	High	
Color	Black			
	Gray			
	White			
	Blue			
	Green			
	Red			
	Yellow			

Worksheet 6a: Matching Materials

List up to three materials that best satisfy all the requirements listed in Worksheet 3.

Part A		
1.	2.	3.

Worksheet 7a: List of Matching Materials

Part B				
Properties	**Value Selected**			**Matching Materials**
Hardness	Low	Medium	High	
Impact Strength	Low	Medium	High	
Rigidity	Low	Medium	High	
Abrasion Resistance	Low	Medium	High	
Optical Clarity	Low	Medium	High	
Chemical Resistance	Low	Medium	High	
Food Compatibility	Low	Medium	High	
Ultraviolet Stability	Low	Medium	High	
Moisture Resistance	Low	Medium	High	
Lubricity	Low	Medium	High	
Electrical Insulation	Low	Medium	High	
Heat Resistance	Low	Medium	High	
Recyclability	Low	Medium	High	
Degradability	Low	Medium	High	
Cost	Low	Medium	High	
Color		Black		
		Gray		
		White		
		Blue		
		Green		
		Red		
		Yellow		

Worksheet 6b: Matching Materials

List up to three materials that best satisfy all the requirements listed in Worksheet 3.

Part B					
1.		2.		3.	

Worksheet 7b: List of Matching Materials

Part C				
Properties	**Value Selected**			**Matching Materials**
Hardness	Low	Medium	High	
Impact Strength	Low	Medium	High	
Rigidity	Low	Medium	High	
Abrasion Resistance	Low	Medium	High	
Optical Clarity	Low	Medium	High	
Chemical Resistance	Low	Medium	High	
Food Compatibility	Low	Medium	High	
Ultraviolet Stability	Low	Medium	High	
Moisture Resistance	Low	Medium	High	
Lubricity	Low	Medium	High	
Electrical Insulation	Low	Medium	High	
Heat Resistance	Low	Medium	High	
Recyclability	Low	Medium	High	
Degradability	Low	Medium	High	
Cost	Low	Medium	High	
Color	Black			
	Gray			
	White			
	Blue			
	Green			
	Red			
	Yellow			

Worksheet 6c: Matching Materials

List up to three materials that best satisfy all the requirements listed in Worksheet 3.

Part C		
1.	2.	3.

Worksheet 7c: List of Matching Materials

SELECTING PROCESSES

Step 8: Define the Form Requirements
Study the problem description, define the form characteristics desired of the product, and list these below in Worksheet 8: List of Form Requirements.

Part A

Part B

Part C

Worksheet 8: List of Form Requirements

Step 9: Determine the Stages by which the Desired Form is Achieved
Determine how the form requirements, described in Worksheet 8: List of Form Requirements, can be achieved as a series of successive stages. List each part and the necessary operations required to achieve the desired form in stages. If there are multiple parts in the product, call them out as well. List this information below in Worksheet 9: Stages of Form Generation.

Stage	What form is achieved in this stage?
Part A	
Part B	
Part C	

Worksheet 9: Stages of Form Generation

Step 10: Determine the Manufacturing Processes that are Relevant

Refer to the materials selections listed previously in Worksheet 7: List of Matching Materials, and the form requirements listed in Worksheet 9: Stages of Form Generation. Apply the procedure described below to determine the processes that are appropriate for each stage of processing each of the selected materials.

- Load MaterialTool and on the Title page click on the button called "**Processes.**" This takes you to the Processes section of MaterialTool.
- Click on any of the process groups buttons: "**Finishing**", "**Forming**", "**Joining**" or "**Machining**."
  This takes you to the selected processes group page. Notice that all major material families (such as Ceramic, Leather, and Wood) are listed in the definition. By clicking on these words, it is possible to determine the processes applicable to each material group.
- Click on the button with the name of an individual process. This takes you to that particular process page. By reading the process definition, design tips, and visual consequences, determine if the process can be used to achieve the desired form in the particular material.
- For a deeper understanding of the process, click on the green button named "**Show principle**."
  List the selected processes below in Worksheet 10: List of Selected Manufacturing Processes.

Step	Process	Description
Part A		
1.		
2.		
3.		
4.		
5.		
Part B		
1.		
2.		
3.		
4.		
5.		
Part C		
1.		
2.		
3.		
4.		
5.		

Worksheet 10: List of Selected Manufacturing Processes

FINAL SELECTION OF MATERIALS & PROCESSES

Step 11: List the Final Selections of Materials and Processes

Refer to the materials selections listed previously in Worksheet 7: List of Matching Materials, and the form requirements listed in Worksheet 10: List of Selected Manufacturing Processes. List the final choices of materials and manufacturing processes below in the Final Report.

FINAL REPORT	
Part A	
Material Selected	
Properties	
Rationale	
Processes Selected	
Operation	
Rationale	
Part B	
Material Selected	
Properties	
Rationale	
Processes Selected	
Operation	
Rationale	
Part C	
Material Selected	
Properties	
Rationale	
Processes Selected	
Operation	
Rationale	

Worksheet 11: Final Report

Unit 19: Lobster Cracker

Problem
The manufacturer has already decided on shape and size
the lobster cracker shown in the picture. He wishes to use only two
materials in the production of this product. Your are asked to
recommend the materials and manufacturing processes that would
allow for cost-effective production of the lobster cracker meeting the
manufacturer's specifications. You are also to offer advice as
to the most suitable way to apply graphics on the product.
Please be specific in your recommendations.

SELECTING MATERIALS

Step 1: Identify the Parts of the Product
The product shown above may be made up of more than one part. Examine the photograph of the product, and identify all the parts of the product. List these in Worksheet 1: List of all Product Parts.

1.	
2.	
3.	
4.	
5.	
6.	
7.	
8.	
9.	
10.	

Worksheet 1: List of all Product Parts

Step 2: List Unique Parts
Some of the parts listed above may be identical parts. Identify the unique parts of the product. Consider mirrored parts, such as a left half and a right half of a product, to be identical parts. List up to three unique parts in Worksheet 2: Unique Product Parts.

Part A	
Part B	
Part C	

Worksheet 2: Unique Product Parts

Step 3: Define the Desired Property Values

First identify what performance characteristics are required for each part that has been identified above. Circle the property values for the material that satisfy the performance requirements for each part in Worksheet 3: Property Values below. Note that only some of the properties will be relevant for each part; leave the rest blank.

Part A							
Hardness	Low		Medium		High		
Impact Strength	Low		Medium		High		
Rigidity	Low		Medium		High		
Abrasion Resistance	Low		Medium		High		
Optical Clarity	Low		Medium		High		
Chemical Resistance	Low		Medium		High		
Food Compatibility	Low		Medium		High		
Ultraviolet Stability	Low		Medium		High		
Moisture Resistance	Low		Medium		High		
Lubricity	Low		Medium		High		
Electrical Insulation	Low		Medium		High		
Heat Resistance	Low		Medium		High		
Recyclability	Low		Medium		High		
Degradability	Low		Medium		High		
Cost	Low		Medium		High		
Color	Black	Gray	White	Blue	Green	Red	Yellow
Part B							
Hardness	Low		Medium		High		
Impact Strength	Low		Medium		High		
Rigidity	Low		Medium		High		
Abrasion Resistance	Low		Medium		High		
Optical Clarity	Low		Medium		High		
Chemical Resistance	Low		Medium		High		
Food Compatibility	Low		Medium		High		
Ultraviolet Stability	Low		Medium		High		
Moisture Resistance	Low		Medium		High		
Lubricity	Low		Medium		High		
Electrical Insulation	Low		Medium		High		
Heat Resistance	Low		Medium		High		
Recyclability	Low		Medium		High		
Degradability	Low		Medium		High		
Cost	Low		Medium		High		
Color	Black	Gray	White	Blue	Green	Red	Yellow
Part C							
Hardness	Low		Medium		High		
Impact Strength	Low		Medium		High		
Rigidity	Low		Medium		High		
Abrasion Resistance	Low		Medium		High		
Optical Clarity	Low		Medium		High		
Chemical Resistance	Low		Medium		High		
Food Compatibility	Low		Medium		High		
Ultraviolet Stability	Low		Medium		High		
Moisture Resistance	Low		Medium		High		
Lubricity	Low		Medium		High		
Electrical Insulation	Low		Medium		High		
Heat Resistance	Low		Medium		High		
Recyclability	Low		Medium		High		
Degradability	Low		Medium		High		
Cost	Low		Medium		High		
Color	Black	Gray	White	Blue	Green	Red	Yellow

Worksheet 3: Property Values

Step 4: Determine Matching Material Families

On the MaterialTool Title page click on the button called "Materials." This takes you to the Materials Section. Click on the large button called "Properties." This takes you to the Materials Properties page. Notice the list of material families listed on the left: Ceramic, Cement, Fiber, and so on. In the right pane you see a list of properties such as Hardness, Impact Strength, Rigidity, and so on. Click on the values Low, Medium, or High to determine Material Families that match the choices from Worksheet 3: Property Values. List the matching families for Part A in Worksheet 4a, the matching families for Part B in Worksheet 4b, and the matching families for Part C in Worksheet 4c.

Properties	Part A			Matching Materials
	Value Selected			
Hardness	Low	Medium	High	
Impact Strength	Low	Medium	High	
Rigidity	Low	Medium	High	
Abrasion Resistance	Low	Medium	High	
Optical Clarity	Low	Medium	High	
Chemical Resistance	Low	Medium	High	
Food Compatibility	Low	Medium	High	
Ultraviolet Stability	Low	Medium	High	
Moisture Resistance	Low	Medium	High	
Lubricity	Low	Medium	High	
Electrical Insulation	Low	Medium	High	
Heat Resistance	Low	Medium	High	
Recyclability	Low	Medium	High	
Degradability	Low	Medium	High	
Cost	Low	Medium	High	
Color	Black			
	Gray			
	White			
	Blue			
	Green			
	Red			
	Yellow			

Worksheet 4a: Matching Material Families

Step 5: Identify the Most Appropriate Material Families

List up to three material families that satisfy all the requirements listed in Worksheet 4a.

Part A		
1.	2.	3.

Worksheet 5a: List of Matching Material Families

Part B				
Properties	**Value Selected**			**Matching Materials**
Hardness	Low	Medium	High	
Impact Strength	Low	Medium	High	
Rigidity	Low	Medium	High	
Abrasion Resistance	Low	Medium	High	
Optical Clarity	Low	Medium	High	
Chemical Resistance	Low	Medium	High	
Food Compatibility	Low	Medium	High	
Ultraviolet Stability	Low	Medium	High	
Moisture Resistance	Low	Medium	High	
Lubricity	Low	Medium	High	
Electrical Insulation	Low	Medium	High	
Heat Resistance	Low	Medium	High	
Recyclability	Low	Medium	High	
Degradability	Low	Medium	High	
Cost	Low	Medium	High	
Color	Black			
	Gray			
	White			
	Blue			
	Green			
	Red			
	Yellow			

Worksheet 4b: Matching Material Families

List up to three material families that satisfy all the requirements listed in Worksheet 4b.

Part B					
1.		2.		3.	

Worksheet 5b: List of Matching Material Families

Part C				
Properties	**Value Selected**			**Matching Materials**
Hardness	Low	Medium	High	
Impact Strength	Low	Medium	High	
Rigidity	Low	Medium	High	
Abrasion Resistance	Low	Medium	High	
Optical Clarity	Low	Medium	High	
Chemical Resistance	Low	Medium	High	
Food Compatibility	Low	Medium	High	
Ultraviolet Stability	Low	Medium	High	
Moisture Resistance	Low	Medium	High	
Lubricity	Low	Medium	High	
Electrical Insulation	Low	Medium	High	
Heat Resistance	Low	Medium	High	
Recyclability	Low	Medium	High	
Degradability	Low	Medium	High	
Cost	Low	Medium	High	
Color	Black			
	Gray			
	White			
	Blue			
	Green			
	Red			
	Yellow			

Worksheet 4c: Matching Material Families

List up to three material families that satisfy all the requirements listed in Worksheet 4c.

Part C		
1.	2.	3.

Worksheet 5c: List of Matching Material Families

Step 6: Determine Matching Materials
Once again, go to the Materials Section and click on the large button called "Properties." This takes you to the main Materials Properties page. Referring to the list you created in Worksheet 3, click the arrow next to the first family. This brings up detailed materials. In the right pane, you see a list of properties such as Hardness, Impact Strength, Rigidity, and so on. Once again select the materials that match the choices in Worksheet 1. List these below in Worksheet 6: Matching Materials. List the matching materials for Part A in Worksheet 6a, the matching materials for Part B in Worksheet 6b, and the matching materials for Part C in Worksheet 6c.

Part A				
Properties	**Value Selected**			**Matching Materials**
Hardness	Low	Medium	High	
Impact Strength	Low	Medium	High	
Rigidity	Low	Medium	High	
Abrasion Resistance	Low	Medium	High	
Optical Clarity	Low	Medium	High	
Chemical Resistance	Low	Medium	High	
Food Compatibility	Low	Medium	High	
Ultraviolet Stability	Low	Medium	High	
Moisture Resistance	Low	Medium	High	
Lubricity	Low	Medium	High	
Electrical Insulation	Low	Medium	High	
Heat Resistance	Low	Medium	High	
Recyclability	Low	Medium	High	
Degradability	Low	Medium	High	
Cost	Low	Medium	High	
Color	Black			
	Gray			
	White			
	Blue			
	Green			
	Red			
	Yellow			

Worksheet 6a: Matching Materials

List up to three materials that best satisfy all the requirements listed in Worksheet 3.

Part A		
1.	2.	3.

Worksheet 7a: List of Matching Materials

Part B				
Properties	**Value Selected**			**Matching Materials**
Hardness	Low	Medium	High	
Impact Strength	Low	Medium	High	
Rigidity	Low	Medium	High	
Abrasion Resistance	Low	Medium	High	
Optical Clarity	Low	Medium	High	
Chemical Resistance	Low	Medium	High	
Food Compatibility	Low	Medium	High	
Ultraviolet Stability	Low	Medium	High	
Moisture Resistance	Low	Medium	High	
Lubricity	Low	Medium	High	
Electrical Insulation	Low	Medium	High	
Heat Resistance	Low	Medium	High	
Recyclability	Low	Medium	High	
Degradability	Low	Medium	High	
Cost	Low	Medium	High	
Color	Black			
	Gray			
	White			
	Blue			
	Green			
	Red			
	Yellow			

Worksheet 6b: Matching Materials

List up to three materials that best satisfy all the requirements listed in Worksheet 3.

Part B		
1.	2.	3.

Worksheet 7b: List of Matching Materials

Part C				
Properties	**Value Selected**			**Matching Materials**
Hardness	Low	Medium	High	
Impact Strength	Low	Medium	High	
Rigidity	Low	Medium	High	
Abrasion Resistance	Low	Medium	High	
Optical Clarity	Low	Medium	High	
Chemical Resistance	Low	Medium	High	
Food Compatibility	Low	Medium	High	
Ultraviolet Stability	Low	Medium	High	
Moisture Resistance	Low	Medium	High	
Lubricity	Low	Medium	High	
Electrical Insulation	Low	Medium	High	
Heat Resistance	Low	Medium	High	
Recyclability	Low	Medium	High	
Degradability	Low	Medium	High	
Cost	Low	Medium	High	
Color	Black			
	Gray			
	White			
	Blue			
	Green			
	Red			
	Yellow			

Worksheet 6c: Matching Materials

List up to three materials that best satisfy all the requirements listed in Worksheet 3.

Part C				
1.		2.		3.	

Worksheet 7c: List of Matching Materials

SELECTING PROCESSES

Step 8: Define the Form Requirements
Study the problem description, define the form characteristics desired of the product, and list these below in Worksheet 8: List of Form Requirements.

Part A
Part B
Part C

Worksheet 8: List of Form Requirements

Step 9: Determine the Stages by which the Desired Form is Achieved
Determine how the form requirements, described in Worksheet 8: List of Form Requirements, can be achieved as a series of successive stages. List each part and the necessary operations required to achieve the desired form in stages. If there are multiple parts in the product, call them out as well. List this information below in Worksheet 9: Stages of Form Generation.

Stage	What form is achieved in this stage?
Part A	
Part B	
Part C	

Worksheet 9: Stages of Form Generation

Step 10: Determine the Manufacturing Processes that are Relevant
Refer to the materials selections listed previously in Worksheet 7: List of Matching Materials, and the form requirements listed in Worksheet 9: Stages of Form Generation. Apply the procedure described below to determine the processes that are appropriate for each stage of processing each of the selected materials.
- Load MaterialTool and on the Title page click on the button called "**Processes.**" This takes you to the Processes section of MaterialTool.
- Click on any of the process groups buttons: "**Finishing**", "**Forming**", "**Joining**" or "**Machining**."
  This takes you to the selected processes group page. Notice that all major material families (such as Ceramic, Leather, and Wood) are listed in the definition. By clicking on these words, it is possible to determine the processes applicable to each material group.
- Click on the button with the name of an individual process. This takes you to that particular process page. By reading the process definition, design tips, and visual consequences, determine if the process can be used to achieve the desired form in the particular material.
- For a deeper understanding of the process, click on the green button named "**Show principle**."
  List the selected processes below in Worksheet 10: List of Selected Manufacturing Processes.

Step	Process	Description
Part A		
1.		
2.		
3.		
4.		
5.		
Part B		
1.		
2.		
3.		
4.		
5.		
Part C		
1.		
2.		
3.		
4.		
5.		

Worksheet 10: List of Selected Manufacturing Processes

FINAL SELECTION OF MATERIALS & PROCESSES

Step 11: List the Final Selections of Materials and Processes
Refer to the materials selections listed previously in Worksheet 7: List of Matching Materials, and the form requirements listed in Worksheet 10: List of Selected Manufacturing Processes. List the final choices of materials and manufacturing processes below in the Final Report.

FINAL REPORT	
Part A	
Material Selected	
Properties	
Rationale	
Processes Selected	
Operation	
Rationale	
Part B	
Material Selected	
Properties	
Rationale	
Processes Selected	
Operation	
Rationale	
Part C	
Material Selected	
Properties	
Rationale	
Processes Selected	
Operation	
Rationale	

Worksheet 11: Final Report

Unit 20: Cement Edge Smoother

Problem
A cement edge smoother is a tool commonly used in construction to smooth edges of walls. It is used both indoors and outdoors.
What materials and manufacturing processes should be used in the production of this tool to ensure its durability and good performance?

SELECTING MATERIALS

Step 1: Identify the Parts of the Product
The product shown above may be made up of more than one part. Examine the photograph of the product, and identify all the parts of the product. List these in Worksheet 1: List of all Product Parts.

1.	
2.	
3.	
4.	
5.	
6.	
7.	
8.	
9.	
10.	

Worksheet 1: List of all Product Parts

Step 2: List Unique Parts
Some of the parts listed above may be identical parts. Identify the unique parts of the product. Consider mirrored parts, such as a left half and a right half of a product, to be identical parts. List up to three unique parts in Worksheet 2: Unique Product Parts.

Part A	
Part B	
Part C	

Worksheet 2: Unique Product Parts

Step 3: Define the Desired Property Values

First identify what performance characteristics are required for each part that has been identified above. Circle the property values for the material that satisfy the performance requirements for each part in Worksheet 3: Property Values below. Note that only some of the properties will be relevant for each part; leave the rest blank.

Part A						
Hardness	Low		Medium		High	
Impact Strength	Low		Medium		High	
Rigidity	Low		Medium		High	
Abrasion Resistance	Low		Medium		High	
Optical Clarity	Low		Medium		High	
Chemical Resistance	Low		Medium		High	
Food Compatibility	Low		Medium		High	
Ultraviolet Stability	Low		Medium		High	
Moisture Resistance	Low		Medium		High	
Lubricity	Low		Medium		High	
Electrical Insulation	Low		Medium		High	
Heat Resistance	Low		Medium		High	
Recyclability	Low		Medium		High	
Degradability	Low		Medium		High	
Cost	Low		Medium		High	
Color	Black	Gray	White	Blue	Green	Red	Yellow

Part B						
Hardness	Low		Medium		High	
Impact Strength	Low		Medium		High	
Rigidity	Low		Medium		High	
Abrasion Resistance	Low		Medium		High	
Optical Clarity	Low		Medium		High	
Chemical Resistance	Low		Medium		High	
Food Compatibility	Low		Medium		High	
Ultraviolet Stability	Low		Medium		High	
Moisture Resistance	Low		Medium		High	
Lubricity	Low		Medium		High	
Electrical Insulation	Low		Medium		High	
Heat Resistance	Low		Medium		High	
Recyclability	Low		Medium		High	
Degradability	Low		Medium		High	
Cost	Low		Medium		High	
Color	Black	Gray	White	Blue	Green	Red	Yellow

Part C						
Hardness	Low		Medium		High	
Impact Strength	Low		Medium		High	
Rigidity	Low		Medium		High	
Abrasion Resistance	Low		Medium		High	
Optical Clarity	Low		Medium		High	
Chemical Resistance	Low		Medium		High	
Food Compatibility	Low		Medium		High	
Ultraviolet Stability	Low		Medium		High	
Moisture Resistance	Low		Medium		High	
Lubricity	Low		Medium		High	
Electrical Insulation	Low		Medium		High	
Heat Resistance	Low		Medium		High	
Recyclability	Low		Medium		High	
Degradability	Low		Medium		High	
Cost	Low		Medium		High	
Color	Black	Gray	White	Blue	Green	Red	Yellow

Worksheet 3: Property Values

Step 4: Determine Matching Material Families

On the MaterialTool Title page click on the button called "Materials." This takes you to the Materials Section. Click on the large button called "Properties." This takes you to the Materials Properties page. Notice the list of material families listed on the left: Ceramic, Cement, Fiber, and so on. In the right pane you see a list of properties such as Hardness, Impact Strength, Rigidity, and so on. Click on the values Low, Medium, or High to determine Material Families that match the choices from Worksheet 3: Property Values. List the matching families for Part A in Worksheet 4a, the matching families for Part B in Worksheet 4b, and the matching families for Part C in Worksheet 4c.

Part A				
Properties	**Value Selected**			**Matching Materials**
Hardness	Low	Medium	High	
Impact Strength	Low	Medium	High	
Rigidity	Low	Medium	High	
Abrasion Resistance	Low	Medium	High	
Optical Clarity	Low	Medium	High	
Chemical Resistance	Low	Medium	High	
Food Compatibility	Low	Medium	High	
Ultraviolet Stability	Low	Medium	High	
Moisture Resistance	Low	Medium	High	
Lubricity	Low	Medium	High	
Electrical Insulation	Low	Medium	High	
Heat Resistance	Low	Medium	High	
Recyclability	Low	Medium	High	
Degradability	Low	Medium	High	
Cost	Low	Medium	High	
Color	Black			
	Gray			
	White			
	Blue			
	Green			
	Red			
	Yellow			

Worksheet 4a: Matching Material Families

Step 5: Identify the Most Appropriate Material Families

List up to three material families that satisfy all the requirements listed in Worksheet 4a.

Part A		
1.	2.	3.

Worksheet 5a: List of Matching Material Families

Part B				
Properties	**Value Selected**			**Matching Materials**
Hardness	Low	Medium	High	
Impact Strength	Low	Medium	High	
Rigidity	Low	Medium	High	
Abrasion Resistance	Low	Medium	High	
Optical Clarity	Low	Medium	High	
Chemical Resistance	Low	Medium	High	
Food Compatibility	Low	Medium	High	
Ultraviolet Stability	Low	Medium	High	
Moisture Resistance	Low	Medium	High	
Lubricity	Low	Medium	High	
Electrical Insulation	Low	Medium	High	
Heat Resistance	Low	Medium	High	
Recyclability	Low	Medium	High	
Degradability	Low	Medium	High	
Cost	Low	Medium	High	
Color	Black			
	Gray			
	White			
	Blue			
	Green			
	Red			
	Yellow			

Worksheet 4b: Matching Material Families

List up to three material families that satisfy all the requirements listed in Worksheet 4b.

Part B		
1.	2.	3.

Worksheet 5b: List of Matching Material Families

Part C				
Properties	**Value Selected**			**Matching Materials**
Hardness	Low	Medium	High	
Impact Strength	Low	Medium	High	
Rigidity	Low	Medium	High	
Abrasion Resistance	Low	Medium	High	
Optical Clarity	Low	Medium	High	
Chemical Resistance	Low	Medium	High	
Food Compatibility	Low	Medium	High	
Ultraviolet Stability	Low	Medium	High	
Moisture Resistance	Low	Medium	High	
Lubricity	Low	Medium	High	
Electrical Insulation	Low	Medium	High	
Heat Resistance	Low	Medium	High	
Recyclability	Low	Medium	High	
Degradability	Low	Medium	High	
Cost	Low	Medium	High	
Color	Black			
	Gray			
	White			
	Blue			
	Green			
	Red			
	Yellow			

Worksheet 4c: Matching Material Families

List up to three material families that satisfy all the requirements listed in Worksheet 4c.

Part C		
1.	2.	3.

Worksheet 5c: List of Matching Material Families

Step 6: Determine Matching Materials
Once again, go to the Materials Section and click on the large button called "Properties." This takes you to the main Materials Properties page. Referring to the list you created in Worksheet 3, click the arrow next to the first family. This brings up detailed materials. In the right pane, you see a list of properties such as Hardness, Impact Strength, Rigidity, and so on. Once again select the materials that match the choices in Worksheet 1. List these below in Worksheet 6: Matching Materials. List the matching materials for Part A in Worksheet 6a, the matching materials for Part B in Worksheet 6b, and the matching materials for Part C in Worksheet 6c.

Properties	Value Selected			Matching Materials
		Part A		
Hardness	Low	Medium	High	
Impact Strength	Low	Medium	High	
Rigidity	Low	Medium	High	
Abrasion Resistance	Low	Medium	High	
Optical Clarity	Low	Medium	High	
Chemical Resistance	Low	Medium	High	
Food Compatibility	Low	Medium	High	
Ultraviolet Stability	Low	Medium	High	
Moisture Resistance	Low	Medium	High	
Lubricity	Low	Medium	High	
Electrical Insulation	Low	Medium	High	
Heat Resistance	Low	Medium	High	
Recyclability	Low	Medium	High	
Degradability	Low	Medium	High	
Cost	Low	Medium	High	
Color		Black		
		Gray		
		White		
		Blue		
		Green		
		Red		
		Yellow		

Worksheet 6a: Matching Materials

List up to three materials that best satisfy all the requirements listed in Worksheet 3.

Part A					
1.		2.		3.	

Worksheet 7a: List of Matching Materials

228

Part B				
Properties	**Value Selected**			**Matching Materials**
Hardness	Low	Medium	High	
Impact Strength	Low	Medium	High	
Rigidity	Low	Medium	High	
Abrasion Resistance	Low	Medium	High	
Optical Clarity	Low	Medium	High	
Chemical Resistance	Low	Medium	High	
Food Compatibility	Low	Medium	High	
Ultraviolet Stability	Low	Medium	High	
Moisture Resistance	Low	Medium	High	
Lubricity	Low	Medium	High	
Electrical Insulation	Low	Medium	High	
Heat Resistance	Low	Medium	High	
Recyclability	Low	Medium	High	
Degradability	Low	Medium	High	
Cost	Low	Medium	High	
Color	Black			
	Gray			
	White			
	Blue			
	Green			
	Red			
	Yellow			

Worksheet 6b: Matching Materials

List up to three materials that best satisfy all the requirements listed in Worksheet 3.

Part B					
1.		2.		3.	

Worksheet 7b: List of Matching Materials

Part C				
Properties	**Value Selected**			**Matching Materials**
Hardness	Low	Medium	High	
Impact Strength	Low	Medium	High	
Rigidity	Low	Medium	High	
Abrasion Resistance	Low	Medium	High	
Optical Clarity	Low	Medium	High	
Chemical Resistance	Low	Medium	High	
Food Compatibility	Low	Medium	High	
Ultraviolet Stability	Low	Medium	High	
Moisture Resistance	Low	Medium	High	
Lubricity	Low	Medium	High	
Electrical Insulation	Low	Medium	High	
Heat Resistance	Low	Medium	High	
Recyclability	Low	Medium	High	
Degradability	Low	Medium	High	
Cost	Low	Medium	High	
Color	Black			
	Gray			
	White			
	Blue			
	Green			
	Red			
	Yellow			

Worksheet 6c: Matching Materials

List up to three materials that best satisfy all the requirements listed in Worksheet 3.

Part C			
1.	2.	3.	

Worksheet 7c: List of Matching Materials

SELECTING PROCESSES

Step 8: Define the Form Requirements
Study the problem description, define the form characteristics desired of the product, and list these below in Worksheet 8: List of Form Requirements.

Part A
Part B
Part C

Worksheet 8: List of Form Requirements

Step 9: Determine the Stages by which the Desired Form is Achieved
Determine how the form requirements, described in Worksheet 8: List of Form Requirements, can be achieved as a series of successive stages. List each part and the necessary operations required to achieve the desired form in stages. If there are multiple parts in the product, call them out as well. List this information below in Worksheet 9: Stages of Form Generation.

Stage	What form is achieved in this stage?
	Part A
	Part B
	Part C

Worksheet 9: Stages of Form Generation

Step 10: Determine the Manufacturing Processes that are Relevant
Refer to the materials selections listed previously in Worksheet 7: List of Matching Materials, and the form requirements listed in Worksheet 9: Stages of Form Generation. Apply the procedure described below to determine the processes that are appropriate for each stage of processing each of the selected materials.

- Load MaterialTool and on the Title page click on the button called "**Processes.**" This takes you to the Processes section of MaterialTool.
- Click on any of the process groups buttons: "**Finishing**", "**Forming**", "**Joining**" or "**Machining**."
  This takes you to the selected processes group page. Notice that all major material families (such as Ceramic, Leather, and Wood) are listed in the definition. By clicking on these words, it is possible to determine the processes applicable to each material group.
- Click on the button with the name of an individual process. This takes you to that particular process page. By reading the process definition, design tips, and visual consequences, determine if the process can be used to achieve the desired form in the particular material.
- For a deeper understanding of the process, click on the green button named "**Show principle**."
  List the selected processes below in Worksheet 10: List of Selected Manufacturing Processes.

Step	Process	Description
Part A		
1.		
2.		
3.		
4.		
5.		
Part B		
1.		
2.		
3.		
4.		
5.		
Part C		
1.		
2.		
3.		
4.		
5.		

Worksheet 10: List of Selected Manufacturing Processes

FINAL SELECTION OF MATERIALS & PROCESSES

Step 11: List the Final Selections of Materials and Processes

Refer to the materials selections listed previously in Worksheet 7: List of Matching Materials, and the form requirements listed in Worksheet 10: List of Selected Manufacturing Processes. List the final choices of materials and manufacturing processes below in the Final Report.

FINAL REPORT	
Part A	
Material Selected	
Properties	
Rationale	
Processes Selected	
Operation	
Rationale	
Part B	
Material Selected	
Properties	
Rationale	
Processes Selected	
Operation	
Rationale	
Part C	
Material Selected	
Properties	
Rationale	
Processes Selected	
Operation	
Rationale	

Worksheet 11: Final Report

SOLUTIONS

Listed below are the solutions for units. Note that a few products have identical or symmetrical parts. The student is encouraged to focus on identifying materials and manufacture of dissimilar parts.
A few products are made up of more than three parts including small components such as screws, rivets, and pins. The student is encouraged to focus on the major parts of each product.

Unit 1:	**Pastry Container**
	Material 1: Polystyrene
	Process 1: Shearing
	Process 2: Vacuum Forming
	Process 3: Trimming

Unit 2:	**Sake Box**
Part A.	Material 1: Pine Wood *(4 sides)*
	Process 1: Circular Sawing
	Process 2: Planning
	Process 3: Milling
	Process 4: Adhesive Bonding
Part B.	Material 1: Circular Sawing *(Bottom)*
	Process 1: Planning
	Process 2: Adhesive Bonding

Unit 3:	**Stirrer**
Part A.	Material 1: Birch Wood
	Process 1: Circular Sawing
	Process 2: Band Sawing
	Process 3: Abrasive Machining
	Process 4: Drilling
	Process 5: Abrasive Finishing

Unit 4:	**Table Cloth**
	Material 1: Cotton
	Process 1: Shearing
	Process 2: Chain Stitching
	Process 3: Flat Seam Stitching
	Process 4: Zig-Zag Stitching

Unit 5:	**Storage Box**
Part A.	Material 1: Galvanized Steel
	Process 1: Shearing
	Process 2: Brake Bending
	Process 3: Riveting
PART B.	Material 1: Galvanized Steel
	Process 1: Shearing
	Process 2: Brake Bending
PART C.	Material 1: Low Carbon Steel *(Rivets)*
	Process 1: Cold Heading

Unit 6:	**Screen Construction Block**
	Material 1: Concrete
	Process 1: Casting

Unit 7:	**Teapot**
Part A.	Material 1: China
	Process 1: Slush Casting (Slip Casting)
	Process 2: Firing
	Process 3: Glazing
Part B.	Material 1: China
	Process 1: Slush Casting (Slip Casting)
	Process 2: Firing
	Process 3: Glazing

Unit 8:   Nail Clipper

Part A.   Material 1: Stainless Steel

Process 1: Blanking

Process 2: Punching

Process 3: Stamping

Part B.   Material 1: Stainless Steel *(2 parts)*

Process 1: Blanking

Process 2: Punching

Process 3: Stamping

Process 4: Grinding

Process 5: Riveting

Part C.   Material 1: Stainless Steel *(Nail File)*

Process 1: Blanking

Process 2: Punching

Process 3: Shaping

Unit 9:   Belt

Part A.   Material 1: Cattle Hide

Process 1: Slitting

Process 2: Punching

Process 3: Stitching

Process 4: Polishing

Part B.   Material 1: Alloy Steel

Process 1: Forging

Process 2: Polishing

Part C.   Material 1: Alloy Steel *(Buckle Tongue)*

Process 1: Shearing

Process 2: Bending

Process 3: Polishing

Unit 10:	**Garden Tool**
Part A.	Material 1: Steel
	Process 1: Shearing
	Process 2: Flexible Die Forming
	Process 3: Forging
	Process 4: Galvanizing
	Process 5: Gas Welding
Part B.	Material 1: Steel
	Process 1: Shearing
	Process 2: Bending
	Process 3: Forging
	Process 4: Galvanizing
Part C.	Material 1: ABS
	Process 1: Injection Molding
	Process 2: Thermo Shrink

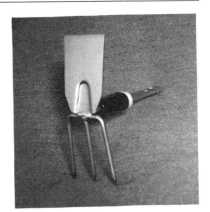

Unit 11:	**Salt & Pepper Shakers**
Part A.	Material 1: Soda-Lime Glass
	Process 1: Blow Molding
Part B.	Material 1: Melamine
	Process 1: Compression Molding

Unit 12:	**Bottle Opener**
Part A.	Material 1: Stainless Steel
	Process 1: Blanking
	Process 2: Forging
	Process 3: Polishing
Part B.	Material 1: ABS
	Process 1: Injection Molding

Unit 13: Garlic Press

Part A. Material 1: Aluminum Die Casting Alloy
Process 1: Die Casting
Process 2: Polishing

Part B. Material 1: Styrene Butadiene
Process 1: Injection Molding

Part C. Material 1: Stainless Steel *(Pin)*
Process 1: Rolling
Process 2: Circular Sawing

Unit 14: Baby Bottle

Part A. Material 1: Latex
Process 1: Dip Molding (Dip Casting)

Part B. Material 1: Polystyrene
Process 1: Injection Molding
(Injection Blow Molding)
Process 2: Screen Printing

Part C. Material 1: ABS
Process 1: Injection Molding

Unit 15: Coffee Pot

Part A. Material 1: Soda-Lime Glass
Process 1: Injection Blow Molding

Part B. Material 1: ABS
Process 1: Injection Molding

Part C. Material 1: Steel
Process 1: Shearing
Process 2: Bending
Process 3: Drilling
Process 4: Threaded Joining

Unit 16: **Clothes Clip**

Part A. Material 1: Pine Wood
Process 1: Circular Sawing
Process 2: Planing
Process 3: Milling

Part B. Material 1: Medium Carbon Steel
Process 1: Bottom Bending

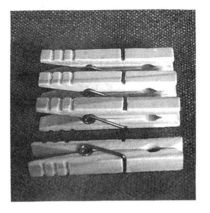

Unit 17: **Grater**

Part A. Material 1: Stainless Steel
Process 1: Shearing
Process 2: Stamping
Process 3: Bottom Bending
Process 4: Punching
Process 5: Seam Locking

Part B. Material 1: ABS
Process 1: Injection Molding

Unit 18: **Book**

Part A. Material 1: Offset Paper
Process 1: Offset Printing
Process 2: Guillotine Shearing
Process 3: Stitching
Process 4: Adhesive Bonding

Unit 19:	**Lobster Cracker**
Part A.	Material 1: Nylon
	Process 1: Injection Molding
Part B.	Material 1: Stainless Steel *(Pin)*
	Process 1: Rolling
	Process 2: Circular Sawing

Unit 20:	**Cement Edger**
Part A.	Material 1: Stainless Steel
	Process 1: Shearing
	Process 2: Blanking
	Process 3: Polishing
Part B	Material 1: Stainless Steel
	Process 1: Shearing
	Process 2: Bottom Bending
	Process 3: Spot Welding
Part C	Material 1: Pine Wood
	Process 1: Circular Sawing
	Process 2: Turning
	Process 3: Milling
	Process 4: Threaded Joining